SURV 9 - Surveillance, Search and Rescue Craft

AA005856

London, United Kingdom
18 April 2018

ISBN: 978-1-5108-8322-2

Printed from e-media with permission by:

Curran Associates, Inc.
57 Morehouse Lane
Red Hook, NY 12571

Some format issues inherent in the e-media version may also appear in this print version.

Printed with permission by Curran Associates, Inc. (2026)

For permission requests, please contact The Royal Institution of Naval Architects at the address below.

The Royal Institution of Naval Architects
8-9 Northumberland Street
London, WC2N 5DA
United Kingdom

Phone: 020 7235 4622
Fax: 020 7259 5912

publications@rina.org.uk

Additional copies of this publication are available from:

Curran Associates, Inc.
57 Morehouse Lane
Red Hook, NY 12571 USA
Phone: 845-758-0400
Fax: 845-758-2633
Email: curran@proceedings.com
Web: www.proceedings.com

CONTENTS

STRUCTURAL ANALYSIS OF COMPOSITE SEARCH AND RESCUE VESSELS UNDER THE NEW BS EN ISO 12215-5

J-B R G Souppez, Southampton Solent University, UK.

SUMMARY

Amongst the new implementations to the forthcoming BS EN ISO 12215-5 features a workboat annex, namely Annex J. This will allow vessels in professional use, such as search and rescue crafts, to be designed under the updated regulation. Moreover, to account for the increasing design and operating speeds since the previous BS EN ISO 12215-5 published in 2008, the scope has been extended beyond 50 knots, while also accounting for higher accelerations. The technical background and practical applications of the new regulation for composite vessels will be presented, highlighting the increased factors of safety adopted to ensure reliability and account for the operating profile of the vessels.

NOMENCLATURE

For the purpose of this paper, the following nomenclature applies, as defined in the ISO 8666 [1] and ISO 12215 [2].

B_C	Chine beam (m) at 0.4 L_{WL}
g	Acceleration due to gravity (m.s^{-2})
$H_{1/3}$	Significant wave height (m)
k_{DYN1}	Dynamic load factor (equivalent to gs)
L_H	Length of the hull (m)
m_{LDC}	Maximum loaded displacement (kg)
t	Thickness (mm)
V_R	Recommended reduced speed (kts)
w	Dry mass of fibre reinforcement (kg.m^{-2})
w_{os}	Dry mass of outer skin reinforcement (kg.m^{-2})
$\beta_{0.4}$	Deadrise angle (°) at 0.4 L_{WL}

1. INTRODUCTION

Four years after the publication of the BS EN ISO 12215-5:2008 [2], the working group 18, part of the ISO technical committee 188, began the work on the revision that will later lead to the latest BS EN ISO 12215:2018 [3] standard. Despite not being intended for commercial vessels, a number of regulations, including the MGN 280 (M) [4] and the Brown Code [5] referred to the ISO standard for the structure of commercial vessels such as pilot boats, without any prior consultation of the working group 18. This therefore motivated the development of a workboat annex that extends the scope of the standard to now include commercial vessels.

The background to the revision and the main modifications from the previous version will be introduced, together with the changes to the scope. Then, the regulatory definition of workboats and inherent categories will be outlined, concluding on the specific requirement for those vessels.

2. BACKGROUND TO THE REVISION

Building on the practical experience of the application of the standard, a number of improvements have been suggested by the industry, and various observations resulting from the use of the standard made, including:

- Large panels were penalised, particularly when made in sandwich. Sandwich structures were further handicapped in terms of attached platting.
- Single curvature was considered, as per Class Regulations; however, for small crafts, accounting for double curvature would be very welcome.
- Vessels featuring a high freeboard appeared overly put at disadvantage compared to low freeboard vessels.
- The applicability of the deflection criterion for sandwich and stiffener was questioned.
- A more advanced analysis method of the quasi-isotropic CSM/WR laminates should be proposed. Furthermore, the simplified analysis for single skin was shown to sometimes give lower requirements than the ply-by-analysis; this was perceived as unfair by the industry as a more advanced analysis method with less uncertainty led to higher structural requirements
- The emergence of new design tools, such as Finite Element Methods (FEM) should be offered as an alternative to the standard.
- Higher accelerations than the 6gs considered previously should be investigated for high speed and light crafts operating in professional use (workboats).

The overall philosophy for the updated standard was to widen the opportunities for more modern structural analysis, however not ruling out the possibility to use simplified methods, better suited to smaller yards. Additionally, the revision aimed to ensure a smooth transition; therefore, it was necessary for vessels passing the 2008 standard to still pass the new one.

3. CHANGES TO THE SCOPE

3.1 MAXIMUM LENGTH

A length of 24m is absolutely critical to define the applicability of the regulatory framework; unfortunately, the definition of 24m is inconsistent. On the one hand, the RCD II [6] and ISO standards are applicable only up to a hull length of 24m. On the other hand, the next regulations (IMO, Class Society, etc…) start at 24m Load Line length

[7], defined as the greatest of 96% of the L_{WL} at 85% of the moulded depth, or the length from the front of the stem to the rudder stock axis. Consequently, vessels with large overhangs would typically be above the 24m hull length, but below the 24m Load Line, thus falling into a regulatory '*no man's land*'.

In order to bridge this regulatory gap, the working group 18 decided to extend the scope of the BS EN ISO 12215 up to 24m load line. It is to be noted that, at present, this has only been adopted for the ISO 12215, and not for other standards or the RCD II. It is however hoped this will provide a precedent that would, in time, lead to a more harmonious definition of 24m across regulatory bodies.

3.2 RACING YACHTS

Although beyond the scope of this paper, the question of racing yachts is worth mentioning. Following the publication of the 2008 version, the ISAF, now World Sailing, made compliance with the 12215 compulsory for offshore races, without prior discussion with the working group 18. While the standard is still not applicable to racing yachts designed for professional racing only, considerations for racing yachts features have been made. This includes correction coefficients for sports and racing crafts, as well as considerations for vessels using water ballast/canting keel.

3.3 WORKBOATS

The increasing recognition of the BS EN ISO 12215-5 by several counties as relevant to commercial vessels, despite the standard clearly not being intended to do so, led the working group 18 to consider the addition of workboats as part of the new version, eventually taking the form of Annex J. This prompted further extension of the scope in terms of accelerations and maximum speeds, to better reflect the mode of operation of commercial vessels.

The definition of workboats and inherent category will be presented in Section 5, while the requirements will be tackled in Section 6.

4 NEW CONSIDERATIONS

In addition to the changes to the scope, a number of new considerations and coefficients have been added [8]; some of the most significant ones are presented in the following subsections.

4.1 APPLICABLE METHODS

To broaden the range of methods available to the industry in analysing the structure, six methods will now be usable to determine the scantlings:

1. Simplified method: based on a simple thickness equation for quasi-isotropic GRP single skin.
2. Enhanced method: ply by ply analysis for quasi-isotropic GRP.
3. Developed method: application of CLT to all FRP structures.
4. Direct test: relying on mechanical testing, primarily intended for FRP.
5. FEM: finite element methods using the ISO design pressures and properties, also mostly aimed at FRP.
6. Drop test: applicable to vessels less than 6m in FRP and non-reinforced plastics.

4.2 ASSESSMENT METHOD FACTOR

As previously stated in Section 2, one of the industry criticism towards the previous version of the standard was that, in certain cases, simpler methods would give lower requirements than more advanced ones. To remedy this issue, and prevent it from happening with the larger number of methods available, an assessment method factor k_{AM} was introduced. The intention being to handicap cruder methods, and promote the use of more advanced ones, as reflected in the values of the coefficient shown in Table 1.

Assessment Method	**Value of k_{AM} for FRP**
Method 1: Simplified	0.90
Method 2: Enhanced	0.95
Method 3: Developed	1
Method 4: Direct Test	1
Method 5: FEM	1
Method 6: Drop Test	n/a

Table 1: *Values of K_{AM}.*

4.3 BOAT BUILDING QUALITY COEFFICIENT

In order to reflect the high impact of the build quality on the final mechanical properties of composite material, a build quality coefficient K_{BB} has been developed. The aim is to reward both the higher manufacturing qualities and higher manufacturing processes, and consequently penalise the mechanical properties for low quality and less advanced manufacturing methods. A summary of the K_{BB} values is presented in Table 2.

Quality	**Builder Characteristics**	**Value of K_{BB}**	
		Hand Laid	**Infused / Pre-Preg**
Tested	Mechanical properties tested and high quality control.	1	1
High	Measured fibre weight fraction and high quality control.	0.95	1
Low	No measurement or checking of fibre weight fraction	0.75	0.8

Table 2: *Values of K_{BB}.*

4.4 LONGITUDINAL PRESSURE DISTRIBUTION FACTOR

The longitudinal pressure distribution coefficient has been modified, following industry feedback, to reduce the requirements in the aft section, but also extending the coefficient beyond the Aft Perpendicular (AP) and Forward Perpendicular (FP). A comparison of the pressure distribution coefficient at accelerations of 6gs and 3gs is depicted in Figure 1.

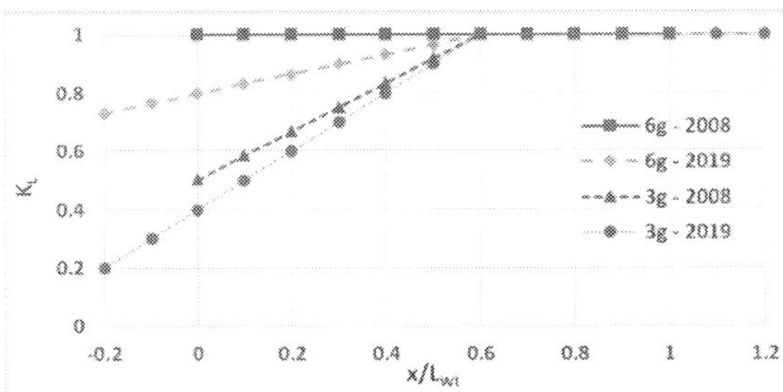

Figure 1: *Changes in the values of K_L.*

While the pressure was assumed to remain constant aft of the AP in the previous version, the new standard will consider a more realistic decrease in pressure over the aft overhang. In the case of workboats, for accelerations between 6 and 8gs, the longitudinal pressure coefficient is to remain at a value of 1 over the entire length.

4.5 ADDITIONAL FACTORS

Among the newly implemented elements of the standard will feature the use of double curvature, and a refined definition for the natural stiffeners on round bilge hull.

On the one hand, while single curvature has long been considered, double curvature was not until now. This is particularly relevant to small crafts, where the curvature can be significant in both directions. The coefficients that will be part of the new standard have primarily been derived from Timoshenko's work [9].

On the other hand, further investigation into the effect of curvature on structures allowed to provide a new definition for natural stiffeners applicable to round bilge hull form. This should prove particularly helpful in reducing the number of stiffeners, particularly in the aft sections of vessels, often deemed 'over structured' by the industry.

5. DEFINITION OF WORKBOATS

5.1 GENERAL

Featured in the inherent normative annex, the definition of workboats distinguishes between charter, light duty and heavy duty crafts. All types are defined as commercial vessels; the distinction then depends on the type of operation and limiting environmental conditions (Beaufort wind speed and significant wave height).

While military crafts and vessels operating in ice conditions are excluded, the intended usage considers a range of boats, from pilot boats and transport vessels for less than 12 passengers to search and rescue crafts.

5.2 CHARTER

Rental and charter vessels do not have any environmental restriction with the exception of the design category conditions. As a commercial vessel, relevant maintenance and survey program are to be implemented. Furthermore, the vessel is expected to be handle with 'good seamanship', and speed reduction when operating above category D should be considered

5.3 LIGHT DUTY

A light duty workboat, is expected to operate in category D, or up to category C restricted to Beaufort 5 and a significant wave height of 1 m. The operating conditions for light duty workboats should not include rough seas, and the comfort of passengers should be paramount, leading to appropriate course and speeds at sea, i.e. strong consideration for seakeeping in order to minimise passenger discomfort. Maintenance and surveying programme shall be undertaken as appropriate, based on the usage and weather conditions experimented.

5.4 HEAVY DUTY

A heavy duty workboat is characterized as operating from the upper end of category C, up to category A, however restricted to Beaufort 9 and 5 m significant wave height. In this particular case, it is assumed that, due to the operating profile of vessels such as search and rescue crafts, the course would not be altered and the speed would not be reduced, and the boat would experience rough seas routinely. Consequently, the 50 knots top speed has been lifted, and accelerations up to 8gs may be considered on the structure; this represents another major change to the scope of the standard. This would obviously require special seating to be provided to the crew in order to remain in full ability to manoeuvre the vessel and be comfortable, as well as imply additional structural requirements. Once again, a suitable maintenance and survey plan shall be implemented.

6. SPECIFIC REQUIREMENTS

6.1 GENERAL

In all three cases, the owner's manual shall provide the appropriate definition of the commercial craft usage conditions, as well the relevant recommendation linked to the specific application. For charter and light duty workboat, this is the only additional requirement incurred by Annex J. Heavy duty workboats however need to satisfy a number of extra criteria.

6.2 HEAVY DUTY

Firstly, as introduced in Section 5.4, accelerations up to 8gs may be considered, and only the first dynamic coefficient criterion k_{DYN1} may be considered. Moreover, the factor of safety for FRP, sandwich core and bulking material has been raised from the standard 2 to 3. The change was deemed relevant by the working group 18, as industry practice is to typically apply a factor of safety of 3 for static loads, and 1.5 for dynamic loads, the latter being rapidly absorbed by the structure. This is achieved by modifying the coefficients in the allowable direct and shear design stress for the materials. In addition, the recommended minimum thickness for single skin and sandwich becomes required, with the addition of a 15% margin (t, w and w_{os} to be multiplied by 1.15). Finally, an equation is provided to suggest the suitable reduction in speed according to the significant wave height experienced, as given equation 1:

$$V_R = \sqrt{\frac{m_{LDC}}{B_C^2} \times \frac{3.125 k_{DYN1}}{\left((50 - \beta_{0.4}) \times \left(\frac{H_{1/3}}{B_C} + 0.084\right)\right)}} \quad (1)$$

In which:

- V_R Recommended reduced speed (kts)
- m_{LDC} Maximum loaded displacement (kg)
- B_C Chine beam (m) at 0.4 L_{WL}
- k_{DYN1} Dynamic load factor
- $\beta_{0.4}$ Deadrise angle (°) at 0.4 L_{WL}
- $H_{1/3}$ Significant wave height (m)

7. CONCLUSIONS

The background to the revision of the BS EN ISO 12215-5 and its impact on the design of composite search and rescue crafts has been presented. Building on the motivation behind the new version, the main changes to the scope have been outlines, including an extension of the applicability up to 24m Load Line, and the inclusion of workboats. Moreover, a number of modern features, such as the applicable methods, assessment method factor, boatbuilding quality factor, longitudinal pressure distribution factor, double curvature and natural stiffeners have been introduced. The definition of workboats and the three subcategories, namely charter, light duty and heavy duty, have been outlined, also detailing the specific requirements for each type. This provides a strong insight into the new regulatory framework and its application to composite vessel in professional use, ahead of the standard being published and becoming compulsory.

8. DISCLAIMER

The views expressed in this paper are those of the author only and do not necessarily reflect those of the ISO/TC188/WG18. All information presented is subject to changes and publication of the final standard.

9. REFERENCES

1. ISO, 'BS EN ISO 8666:2016 – Small Craft – Principal Data', *International Organisation for Standardization*, 2016.

2. ISO, 'BS EN ISO 12215-5:2008+A1:2014 - Small Craft – Hull Construction and Scantlings. Part 5: Design Pressures for Monohulls, Design Stresses, Scantlings Determination', *International Organisation for Standardization*, 2014.

3. ISO, 'BS EN ISO 12215-5:2018 - Small Craft – Hull Construction and Scantlings. Part 5: Design Pressures for Monohulls, Design Stresses, Scantlings Determination', *International Organisation for Standardization*, 2019 (expected).

4. MARITIME AND COASTGUARD AGENCY, 'MGN 280 (M) Small Vessels in Commercial Use for Sport or Pleasure, Workboats and Pilot Boats - Alternative Construction Standards', *Maritime and Coastguard Agency*, 2004.

5. MARITIME AND COASTGUARD AGENCY, 'Code of Practice for the Safety of Small Workboats and Pilot Boats (brown)', *Maritime and Coastguard Agency*, 2001.

6. EUROPEAN PARLIAMENT, 'Directive 2013/53/EU on Recreational Craft and Personal Watercraft', *Official Journal of the European Union*, 2013.

7. INTERNATIONAL MARITIME ORGANIZATION, 'International Convention on Load Lines, International Maritime Organization, 2003.

8. SOUPPEZ, J.-B. R. G. and RIDLEY, J., 'The Revisions of the BS EN ISO 12215', *Marine Sector Showcase – Composite UK*, 2017.

9. TIMOSHENKO, S. P., 'Theory of Plates and Shells', *McGraw-Hill*, Second edition, 1959.

10. AUTHORS BIOGRAPHY

Jean-Baptiste R. G. Souppez holds the position of Senior Lecturer in Yacht Design and Composite Engineering at Southampton Solent University, and contributes to the European Master EMship+ as a Visiting Professor and Research Supervisor. He is also the UK Principal Expert in Small Craft Structures (member of the GME/33 and ISO/TC188/WG18) in charge of representing the interests of the British Marine Industry in the development of international structural regulations (BS EN ISO 12215).

DESIGN VERIFICATION FRAMEWORK FOR SEARCH AND RESCUE CRAFT

H J Phillips, **M Pitman** and **M Boulton**, Royal National Lifeboat Institution, UK
H.E van der Molen, Koninklijke Nederlandse Redding Maatschappij, Netherlands
K Stromenger, K&S Maritiem Technisch Bureau, Netherlands

SUMMARY

Due to the specific requirements and challenges associated with search and rescue craft, it can be difficult to complete designs of such craft in accordance with most codes of practice, regulations and classification society rules.

In 2014, a project (known as SHARK) was set up between the Royal National Lifeboat Institution (RNLI) and Koninklijke Nederlandse Redding Maatschappij (KNRM). The project deliverable was a process that can be followed to provide assurance and design verification of lifeboats, together with a framework based on the Failure Modes, Effects, Criticality and Assessment (FMECA) approach. This can be used as part of the initial design process with the approach also being transferrable to other craft or equipment.

This paper describes the process followed by the SHARK project and the project deliverables. Furthermore, the paper will elaborate on the framework methodology, the verification tool and the associated documentation.

1. INTRODUCTION

In 2014 the Royal Dutch Lifeboat Services (KNRM) launched its prototype lifeboat Nh1816, being a revolutionary design in many aspects. Although the vessel was built under Lloyd's Register classification, it was KNRM's ambition to verify the design of Nh1816 and to translate it into the actual prototype vessel.

A market survey of commercial parties with experience of lifeboat design verification resulted in very limited results. It also became apparent that no format existed by which the desired design verification assurance process could be carried out. KNRM therefore approached the RNLI and requested a design review of its prototype Nh1816 lifeboat. The RNLI was established in 1824 and is coincidently the same year as that of KNRM. The organisation has nearly 200 years' experience of designing and operating lifeboats and at the time, was in the process of developing what became the Shannon class lifeboat. It was, therefore, an ideal time to review the design approaches taken by both organisations of their respective new classes. This request became the start of a two-year collaborative project known as 'SHARK' (**S**hared **HA**ZOP/FMECA **A**ssurance, **R**NLI, **K**NRM).

Operating lifeboats carries unique challenges; All Weather lifeboats must be designed to be able to operate in all kinds of situations, many of which are extremely dangerous. In addition, RNLI and KNRM lifeboat crews are comprised mainly of excellently trained volunteers from all walks of life, many of whom do not have previous maritime experience. Lifeboat institutions are therefore challenged to design their boats such that they are safe to be operated by volunteers and in all conditions. It is therefore not 'just' the lifeboat design that is critical but also its maintainability and simplicity for crews to operate. It is vital that the volunteers are able to interact with all systems on board with as little inherent possibility for error as possible. Consequently, it is perhaps not surprising that although there are some examples, very little experience of this type of lifeboat paradigm exists within third parties.

The RNLI/KRNM project team have produced a framework, SHARK-DAT (Design Assurance Tool). The lifeboat design verification methodology has been broken down into three main elements; 1) actual design with design implications in respect of 2) maintenance and 3) required training to be able to operate the designed system.

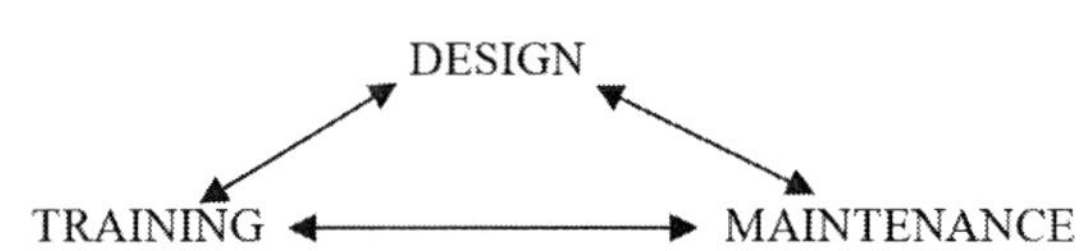

The inter-dependencies of and balance between these three aspects have become a unique feature of the SHARK-DAT framework. A standard FMECA as described in section 4.1 is restricted to systems evaluation only and the HAZOP tool discussed in section 4.2 focusses on the human - system interface. The SHARK-DAT combines both these approaches into one framework. This enables the design team to continuously assess the trade-offs between the three aspects.

Secondly the SHARK-DAT framework can be used to carry out design verification of existing lifeboats, and carries the possibility to compare lifeboats of different designs, even if they are have completely different features. This has been achieved by breaking down a design into sub systems or components and subsequently assessing these sub systems using the framework.

Thirdly, the SHARK-DAT can be used for keeping track of design modifications throughout the lifecycle of the craft and assessing the implications on risk, training and maintenance. For example, if a particular defect were to consistently re-occur such that the reliability of a particular lifeboat was unduly impacted, the proposed changes to rectify the problem must first be passed through an updated SHARK-DAT assessment. This is to capture the effect on the original system to ensure that the most appropriate design change is carried out that not only improves the reliability but such that the associated risks are also acceptable.

Although the SHARK-DAT has been designed by two lifeboat institutions the result is a generic framework that is not just suitable for design assurance of lifeboats, but for any vessel and potentially for any equipment consisting of a design with a requirement for human interaction and that can be broken down into components or sub systems.

2. FLOWCHART – OVERVIEW

Initially, to generate some structure the project team took time to map out the design process from initial requirements to the completed deliverables. The flow chart created guides the user through this design process and is described in more detail in section 5. Once completed the team then had a framework within which the SHARK – DAT tool could be created. Some of the key considerations that underpin the flowchart are discussed below.

The specific systems design procedures were omitted because each design/build authority would have their own existing methods or processes.

The structure created is based around the FMECA and HAZOP approach to design. However, if a boat/ship or system is designed to a known standard, then this approach can be bypassed by the existing rules/guidelines provided by the regulatory authority, provided that it is fully applicable to the scenario being considered.

The process highlights the importance of mitigating any determined risks by reviewing the design of the systems in the first instance. This should ensure a fundamentally safer system prior to the use of training or maintenance procedures as mitigations.

3. EXPLANATION OF ASSURANCE TOOL

The design assurance tool is a semi-automated process that incorporates a set of tools allowing the user to assess a proposed design in a logical and systematic manner. This assessment could be a validation against a design standard or code (where one exists) or to support an organisation's in house safety case. In addition, data entry is recorded such that the design can be modified as necessary and any required mitigating actions captured. The tool can therefore be used and updated throughout the design and operational life of the vessel. Any 'Through-Life' modifications or changes to the vessel should also be passed through this process.

Summary reports can be generated to indicate the quantity of each risk category as described in more detail below. Charts can then be created to provide a highly visual representation of the risks.

It is also possible to compare multiple proposed or existing designs to compare the different approaches taken. Furthermore, the combined reporting for both projects is possible thus highlighting the risk priorities associated with each vessel. This is useful if only limited resource is available such that the highest risks could be addressed as the priority with the others being scheduled in.

4. ASSESSMENT PROCESSES

4.1 FMECA

A FMECA, often referred to as Failure Modes Effect Analysis (FMEA) initially involves a functional assessment of the system in question. This process can be assisted by first generating a diagrammatic functional or system breakdown and defining 'systems', 'sub-systems' and 'components' in addition to determining identification numbers ideally using the SFI code [1]. These items can represent not only physical equipment, but also the associated procedures and the surrounding environment. Functional flow diagrams can be created if considered useful for understanding the relationships between items in a system. Each specified 'system', 'sub-system' or 'component' must be explored and its significant features and functions documented. By carrying out both procedural and equipment assessments, a certain amount of overlap occurs between systems, helping to reduce the chance that any item will be overlooked. [2,3]

The breakdown of a system defines to a large extent the level of ensuing safety assessment. However there is no 'correct' level: engineering judgement must be used to balance the time taken to carry out the safety assessment with the results achieved. In many cases an appropriate level will automatically be reached and acceptable detail of the safety information recorded.

The functional analysis provides a systematic structure for investigating possible failures. A methodical approach involving relevant design, maintenance and operational personnel can then be used to generate the information required by the FMECA.

Each item specified in the functional analysis must be studied in turn, with the initial aim of identifying all possible failure modes that could occur. A failure mode is a way in which an item can fail, and each failure mode must be given a separate row in the FMECA table. For each failure mode, likely immediate causes must be generated. The effects of each failure mode must also be determined. The aim is to determine the overall effect of the failure at the highest level: these final effects are named 'end events' and must be determined at the outset of the FMECA. It is important to ensure that the FMECA is started at the overall system level (e.g. full sea water system).

Local effects can be described as all the effects that occur as a result of the failure that are prior to the end event. The appropriate end events can then be simply chosen from the list. In addition, a qualitative measure of the severity of each end event must also be determined at the outset.

By looking at causes and effects associated with each identified failure mode, possible failure paths ('a combination of basic events which occur together to produce an end event') are being determined. The failure path leading to a specified end event is also known as a 'hazard' e.g. where a risk of injury exists. In order to facilitate the identification of hazards, key words can be used during the session to 'trigger' thought processes. A list of key words that normally form the basis of a standard HAZOP (HAZard and OPerability) assessment can be used. This is covered in more detail in section 4.2 below.

The probability of each identified failure path occurring can be judged using accepted qualitative criteria. The probability is the likelihood that the failure mode will occur **and** lead to the end event i.e. the probability of the path, and not just the failure mode in isolation. After the initial design assessment has been completed, it is beneficial to review it regularly, particularly with regard to probability. Actual defect and incident data can greatly assist in fine tuning probabilities. Any inconsistencies should be rectified as this could have an effect on the recorded level of risk. Section 4.3 describes probabilities, severities and end events in more detail.

Risk is a combination of the probability (likelihood) of a failure path occurring and the severity of consequence of the associated end event. Using a risk matrix, a measure of risk (or criticality) can be associated to each identified failure path. Section 4.4 describes this in more detail.

A failure mitigation process must be carried out whereby risks considered unacceptable can be mitigated to a level of at least 'as low as reasonably practicable' (ALARP). It is likely that the mitigations will reduce the probability rather than the severity so the post action mitigation will be 'End 0' (i.e. the same as before) but with a reduced probability.

4.2 HAZOP

4.2 (a) Description

HAZOP (Hazard & Operability Analysis) is a systematic hazard identification tool which prompts the thinking of the risk assessment team, through the use of sets of keywords, to identify how system users could either induce failures or encounter hazards that may pose a risk of injury or damage [2].

This approach is useful when reviewing hazards that can be difficult to assess with other tools, particularly where the hazard is rooted in human performance and behaviours.
The HAZOP relies heavily on the knowledge and experience of subject matter experts to assess the behaviours and actions of typical users in their interactions with the vessel systems.

A maritime search and rescue asset is ineffective without the people who operate and maintain it. However the actions of those people are sometimes not fully accounted for in the design of the vessel.

Many safety assessment tools (such as FMECA) used in the design of the vessel systems are good for identifying how equipment and systems can fail but they are often not good at capturing operability issues that may arise when operators or maintainers carry out an activity on those systems and equipment.

This potentially leaves a gap in the risk assessment process where the design team believe that the design is good and robust, but the 'people' interactions remain unassessed with the potential for there to be significant risks that only come to light when the vessel is tested or operated and when mitigations are more difficult and expensive to apply.

Additionally, the HAZOP process can identify more detailed user-centred mitigations that can be passed into the operator and maintainer training programmes to ensure that people are competent to use the vessel.

Where people are involved in an operation, it may be worthwhile conducting a HAZOP to identify the operability issues, assess the risks and to identify mitigations to reduce the risk. As for mitigating risks identified using a FMECA approach, these should ideally be design-based.

It would not normally be necessary to conduct both a FMECA and a HAZOP for every system, as there could be a degree of duplication or nugatory work involved. However, for key systems and operations where people have a significant role to play, then a HAZOP assessment should be undertaken.

The HAZOP follows a standard framework, which the assessment team must be familiar with. Therefore training in the system may be required before any assessment work is undertaken.

The selection of the assessment team is fundamental to a successful outcome. Typically the following would be required:

- HAZOP leader – a person familiar with the process and capable of leading the team.
- System designer(s) – who understand the design, particularly the inputs and outputs of the systems concerned.
- Users – Operators and or maintainers (as appropriate) who have experience and background in the use and maintenance of similar equipment and systems.
- A 'wild card' – it can often be useful to involve a person who has limited direct knowledge of the design being reviewed but who has groundings either in general engineering and/or another relevant discipline and who is comfortable challenging some of the debate.
- Administrator – it is usually beneficial to utilise the skills of an administrator to correctly and efficiently capture the information provided by the team

The same definitions of probability, severity and end events are used as for the FMECA and are described in section 4.3 below.

4.2 (b) HAZOP Keywords

The SHARK-DAT tool is pre-set with a standard set of key words that capture typical operability issues. Assessment teams should satisfy themselves that these keywords are appropriate for their use, and they may supplement them with additional keywords if required. The pre-set words include considerations such as early and late/delay, more and less as well as open and close.

For the HAZOP to be effective the assessment team should follow the process of applying each keyword to each aspect of the design/operation, considering the failures/hazards that keyword could result in. Each failure or hazard can then be captured and assessed, with accompanying mitigations noted if necessary.

4.2 (c) HAZOP 'Mindset'

The key when conducting the assessment is to think like a real end user. It should be remembered that users are people, their behaviour is not always predictable and they may not always follow the manual; they may sometimes take positive decisions not to and take short-cuts instead, especially when put under pressure.

The team should consider real life behaviours, tiredness and complacency in their assessments of end user behaviour. The same applies to the team undertaking the HAZOP and FMECA assessment itself. As with all repetitive work, the assessments can become monotonous and it can be very tempting to omit certain words or sections from the process to save time. However, this should only be done by a competent team who have a good understanding and can comprehend the implications. Any such deviations should be recorded for future reference. Consistency, determination, regular breaks and good humour are all key factors in undertaking a successful assessment.

4.3 PROBABILITIES, SEVERITIES AND END EVENTS

To carry out an assessment, it is important to define the probabilities, severities and end events. Those listed below are related to design FMECAs & HAZOPs. If alternative through-life assessments are required (e.g. for a Production or Disposal assessments) then appropriate definitions may need to be prepared and agreed in advance of an assessment.

4.3 (a) Probability Definitions

Code	PROBABILITY	Definition
F	***Frequent***	Likely to occur often during the operational life of a particular craft.
RP	***Reasonably probable***	Unlikely to occur often but may occur several times during the total operational life of a particular craft.
R	***Remote***	Unlikely to occur to every craft but may occur to a few craft of a type over the total operational life of a number of craft of the same type.
ER	*Extremely remote*	Unlikely to occur when considering the total operational life of a number of craft of the type, but nevertheless should be considered as being possible.
EI	*Extremely improbable*	An event that is so extremely remote that it should not be considered as possible to occur.

Table 1: Probability Definitions

4.3 (b) Severity Definitions

Code	SEVERITY	Definition
MI	***Minor***	An event or failure which can be readily compensated for by the crew A small increase in operational duties or in the difficulty of performing duties. A moderate degradation in operational performance. A slight modification of the permissible operating conditions.
MA	***Major***	A significant increase in operational duties or in the difficulty of performing those duties: but not beyond their capability provided another major effect doesn't occur simultaneously. A significant degradation in operational performance. A significant modification of the permissible operating conditions but will not preclude a safe mission.
H	***Hazardous***	A dangerous increase in operational duties or in the difficulty of performing those duties: crew cannot be expected to cope. A dangerous degradation in operational performance and strength of the rig. Marginal conditions for crew. Injury to crew or public. An essential need for outside assistance.
C	***Catastrophic***	Crew fatality. Public fatality. Loss of the boat.

Table 2: Severity Definitions

4.3 (c) End Events

The following list demonstrates how end events are created prior to commencement of the FMECA/HAZOP. Each end event is coded and a severity associated. Appropriate end events of this form must be created for each FMECA and HAZOP.

Code	END EVENT	Details	Severity	
End 0 (Only in Post Action Mitigation)	No change or impact	The system has returned to its original status with the associated severity	Not applicable	-
End 1	Minor personnel injury	Small cuts and bruises Anything that can be treated by minor first aid	Minor	**MI**
End 2	Increased crew effort	Equipment failure on boat but not mission threatening – can be readily compensated for by the crew	Minor	**MI**
End 3	Impact on the Environment	Undesirable impact on environment Discharge into the sea Noise borne pollution Pollution	Major	**MA**
End 4	Boat off service. Cannot embark on service Taken off station	Boat cannot be sent to sea Boat cannot be recovered Boat off service due to trailer fault Boat off service due to crew fault Boat off service due to tractor fault Boat off service due to boat fault	Major	**MA**
End 5	Reduced operational effectiveness	Significant increase in operational duties – can be compensated for by increased crew effort: Reduced performance, handling & engine driven services Increased risk of fire Controllable ingress of water Reduced communications Reduced crew effectiveness	Major	**MA**
End 6	Major personnel injury	Requires first aid and possible trip to casualty Time off work required	Major	**MA**
End 7	Greatly reduced operational effectiveness	Dangerous increase in operational duties – external assistance may be required: Loss of all propulsive or electrical power Total loss of communications Man overboard Reduced stability or buoyancy Capsize Controllable fire	Hazardous	**H**
End 8	Hospitalisation	Head injury, Broken limbs, Major cuts	Hazardous	**H**
End 9	Death or disability	Permanent or life-threatening injury of boat crew, shore crew, public or survivors: Loss of limbs Death	Catastrophic	**C**
End 10	Loss of boat	Uncontrollable fire Boat sinks Boat won't right after capsize Separation from boat	Catastrophic	**C**

Table 3: End Event Definitions

4.4 RISK MATRIX

The combined Probabilities and Severities from the sections above can then be collated into a risk matrix as shown in Table 4. These then provide the risk category as shown in Table 5. The risk categories then describe the required actions. Further details of this are shown on the Flowchart in section 5.

		Severity			
		Minor	**Major**	**Hazardous**	**Catastrophic**
Probability	**Frequent**	A1	U	U	U
	Reasonably Probable	A2	A1	U	U
	Remote	N	A2	A1	U
	Extremely Remote	N	N	A2	A1
	Extremely Improbable	N	N	N	A2

UNACCEPTABLE — ALARP 2 — ALARP 1 — NEGLIGIBLE

Table 4: Risk Matrix

Code		Definition
U	**Unacceptable**	Risk cannot be justified on any grounds
A1	**ALARP 1**	Tolerable only if risk reduction is impractical or if penalties disproportionate to the improvement gained
A2	**ALARP 2**	Tolerable if penalties of reduction would exceed the improvement gained
N	**Negligible**	No need for detailed working to demonstrate ALARP

Table 5: Risk Categories

5. FLOWCHART – STEP BY STEP

As mentioned previously the process of determining the functional and performance requirements of a new vessel and the specifications chosen was deemed out of scope for this project. These decisions are typically determined by the individual clients, designers and builders.

Based on the initial design proposals, the vessel's systems should be split at the highest levels possible. The flow chart can then be used on a 'system by system' basis such that each can be considered separately. It is assumed at this stage that the concept design is developed to a system level with proposed major components having been identified.

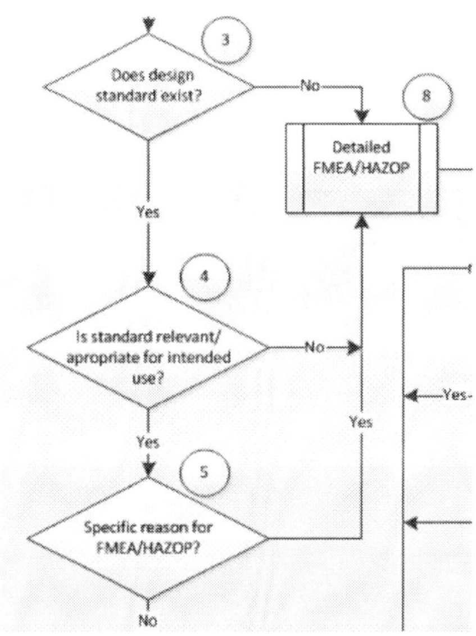

Figure 1: Initial phase

The first main section of the SHARK-DAT is to determine the approach required for the specific system (or whole vessel). This establishes whether or not an FMECA or HAZOP needs to be carried out as shown in Figure 1, Box 3, or if a relevant standard exists (E.g. an ISO, DIN, Classification society rules) to ensure the system is designed to conform. However, it may be required and/or good practice to also include a specific risk based approach (Box 8). It will also be necessary to take a risk based approach if the standard is not followed in its entirety and/or is not entirely relevant (Box 4). The risk based approach requires the completion of a FMECA as a minimum.

In addition, if people are involved in the system process then a HAZOP must also be carried out. There may also be a specific reason for carrying out a FMECA or HAZOP (Box 5). The SHARK-DAT provides a format for carrying out a FMECA and a HAZOP.

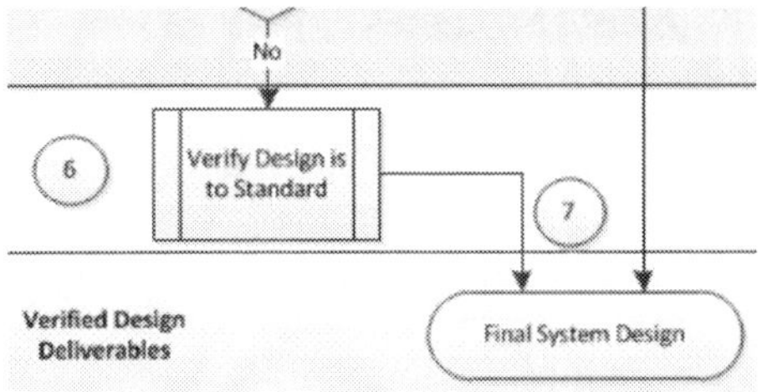

Figure 2: Verification Step if Design Standard used

If the design conforms to an existing standard then it should be verified (Box 6) by a third party (this could be done internally or by an external classification society) prior to development into a final system design (Box 7).

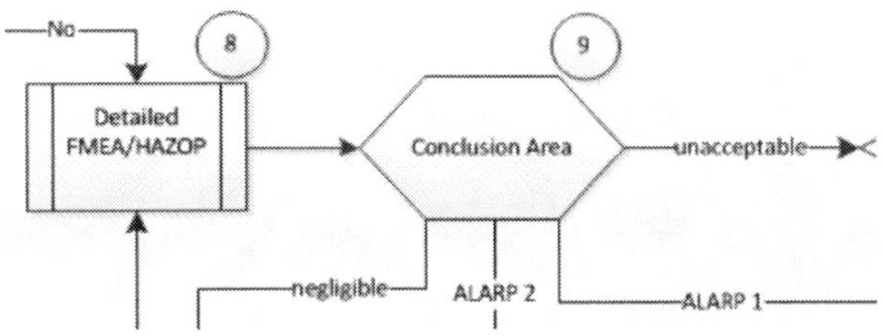

Figure 3: FMECA/HAZOP Conclusion Area

The result of the FMECA and/or HAZOP (using the SHARK-DAT tool) will provide the user with the level of risk calculated in the Conclusion Area (Box 9). The determined level of risk will be Negligible, ALARP 1, ALARP 2 or Unacceptable each providing the user with the next path to follow.

If an outcome is **Negligible** then the final system design can be progressed as shown below in Figure 4.

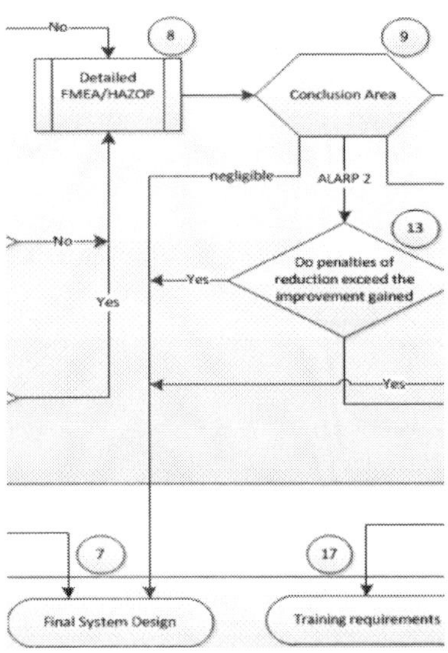

Figure 4: Negligible Conclusion route

For risks other than Negligible, further actions are required. The section of the flowchart in Figure 5 captures these requirements:

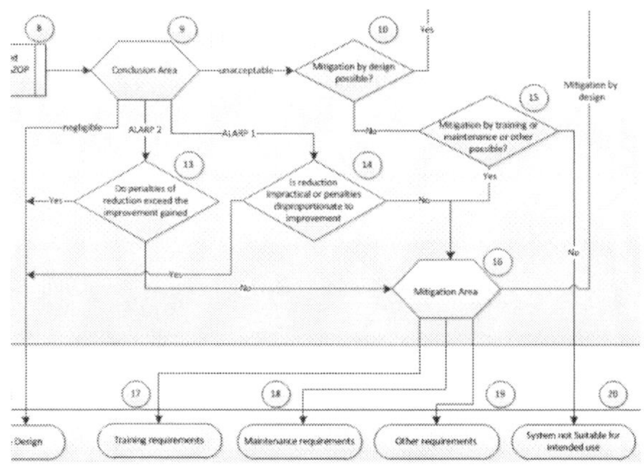

Figure 5: ALARP 2, ALARP 1 and Unacceptable Conclusion routes

In the case of risks that have been categorised as **ALARP 2**, it should be assessed and questioned if the penalties of reducing the risk would exceed the improvement gained or not (Box 13). The penalties would include aspects such as practical implementation, ease of use, maintenance implications and cost. If all risk reduction methods result in penalties that would exceed the improvement gained then these should be recorded on the SHARK-DAT with comments in the 'Action' column and the final system design can be progressed. If risk reduction methods are practical then the Mitigation Area (Box 16) is entered on the flowchart.

In the case of risks that have been categorised as **ALARP 1**, these are only tolerable if the risk reduction is impractical (e.g. materials required are not readily available) or if penalties are disproportionate to the improvement gained (Box 14). If the latter, then the risk reduction options must be recorded on the SHARK-DAT and the final system design progressed. If risk reduction methods are practical then the Mitigation Area (Box 16) is entered on the flowchart.

For risks categorised as **Unacceptable**, then mitigation by design is required (Box 10). If this is not possible then mitigation via another means is required (Box 15). If no other mitigating actions are possible and/or do not reduce the level of risk to ALARP 1 or lower, then the system is not suitable for intended use (Box 20) as shown in Figure 6.

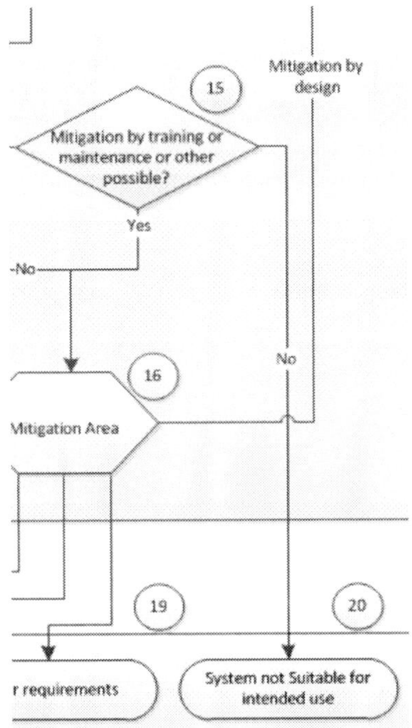

Figure 6: System not suitable for intended use route

The Mitigation area (Box 16) is the stage at which decisions are made to determine the most appropriate methods of reducing the risk. Mitigation through modification of the initial design (Box 11) should always be the preferred first method over and above any other. If this is not possible due to design constraints then suitable mitigation through a combination of training (Box 17), maintenance (Box 18) or other (Box 19) must be specified and documented e.g. through standard operating procedures and planned maintenance documentation.

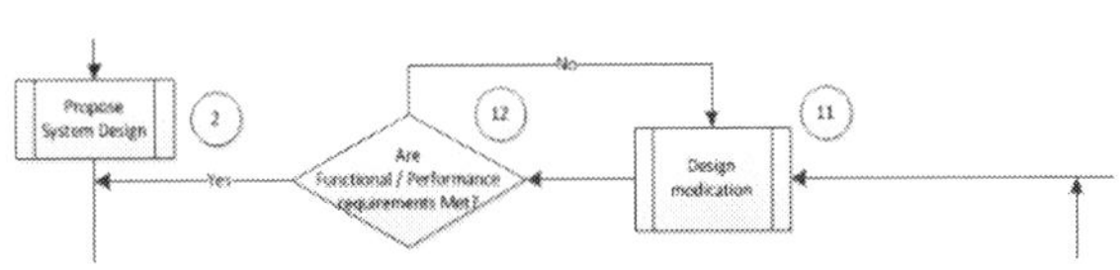

Figure 7: Design Modification Route

Any mitigation by design will result in a design modification (Box 11). At this stage the design needs to be checked against the functional and performance requirements to ensure they are still being met (Box 12). This would be an iterative process until such time that this is the case. When the requirements are met, then the system is re-assessed using the SHARK-DAT.

It is possible that there may be occasions when it is not possible to meet the performance and functional requirement using a safely designed system. At this point, it would be necessary to review the requirements with the customer/operator and budget holder.

On completion of the design, the final set of deliverables including the final system design drawings and associated documentation can be produced. This should include all production, training and maintenance requirements including those specific items that were required as mitigating actions from the mitigation area of the flowchart.

For completeness, the full flowchart is shown in Figure 8 at the end of this paper.

6. CONCLUSIONS

The project has created an extremely useful tool based on a systematic and consistent methodology and in the context of a broader assurance framework. The method encourages the users to consider subsequent maintenance and training requirements at the outset and also how these link together to produce safe and effective designs.

The approach also highlights the importance of considering the end users/operators and capturing the associated risks and required mitigating actions.

The approach described in this paper can be used for any equipment that comprises multiple systems, although the system and sub system descriptions would need to be amended to suit. It is also possible not only to compare designs of different classes but also the approaches taken to ensure consistency.

In addition, the value of working as a collaborative team from two organisations has been highlighted. Involving a representative of a similar organisation to assess a vessel's design (e.g. as a 'wild card) has proven to be extremely beneficial. Project SHARK took advantage of the experiences and expertise of representatives of the KNRM and RNLI to not only develop a working tool but also to provide a high level of assurance of each organisations' respective new classes of lifeboat.

The framework and tool are freely available via the International Marine Rescue Federation (IMRF) website. The authors hope that by sharing the project deliverables, they can contribute in some way towards improving maritime safety.

7. ACKNOWLEDGEMENTS

The authors wish to thank the following:

- The IMRF for making the deliverables from this project available on their website
- Chris Walters, RNLI Trustee and Chair of the RNLI Technical Committee for helping the project in the early stages
- Neil Chaplin, RNLI Head of Asset Management and Andrew Tate, Head of In Service Support for their help and support throughout the project

- Bernard Klaassen of "Effectief Office" for coding framework
- Jeremy Dunn, RNLI Studio Manager for writing the e-learning package

8. REFERENCES

1. The SFI Group System was first released in 1972 as the result of a research project undertaken by the Ship Research Institute of Norway (SFI: Skipsteknisk Forskningsinstitutt).

2. BIRMINGHAM, R., SEN, P., CAIN, C., & CRIPPS, R. M., 'Development and Implementation of a Design for Safety Procedure for search and Rescue Craft', *Journal of Engineering Design, Volume 11, issue 1, Pp 55-78*, 2000.

3. KUO, C., 'Safety Management and its Maritime Application', *The Nautical Institute ISBN 1 870077 830*, 2007.

9. AUTHORS BIOGRAPHY

Holly Phillips holds the current position of Principal Naval Architect at the Royal National Lifeboat Institution. She forms part of the RNLI's Design Authority working group and manages the naval architecture team responsible for the structural and stability aspects of the RNLI's fleet of all weather and inshore lifeboats, lifeguard equipment and other assets. Her previous experience includes that of Project Manager responsible for the design and development of the E-Class lifeboats for the River Thames.

Martyn Pitman holds the current position of Senior Engineer (Trials) at the Royal National Lifeboat Institution. He forms part of the RNLI's Internal Construction Support Cell as subject matter expert for the RNLI's Shannon class lifeboat machinery and mechanical systems. He has over 20 years' experience in the Marine Mechanical and Machinist disciplines, His recent experience includes that of Trials Engineer responsible for the Sea Going trials, testing and management of any required repairs or modifications during the development of the Shannon Class.

Mark Boulton holds the current position of Principal Engineer (Electrical) at the Royal National Lifeboat Institution. He forms part of the RNLI's Internal Construction Support Cell as subject matter expert for the RNLI's Shannon class lifeboat Electrical and Electronic systems. He has over 30 years' experience in the Marine Electrical discipline, starting out as a marine electrician in a shipbuilding yard on the Isle of Wight.

Hans van der Molen holds the current position of Superintendent at the Koninklijke Nederlandse Redding Maatschappij (KNRM). His responsibilities include amongst other, technical innovations. Hans has 30 years' experience with KNRM which covers all technical areas of building and maintaining lifeboats and related technology/equipment.

Klaus Stromenger holds the current position of owner at K&S Maritiem Technisch Bureau. Klaus is a naval architect and has been responsible for the design of various classic passenger carrying sailing ships up to 70m. K&S is also well known for the design of cranes on luxury yachts.

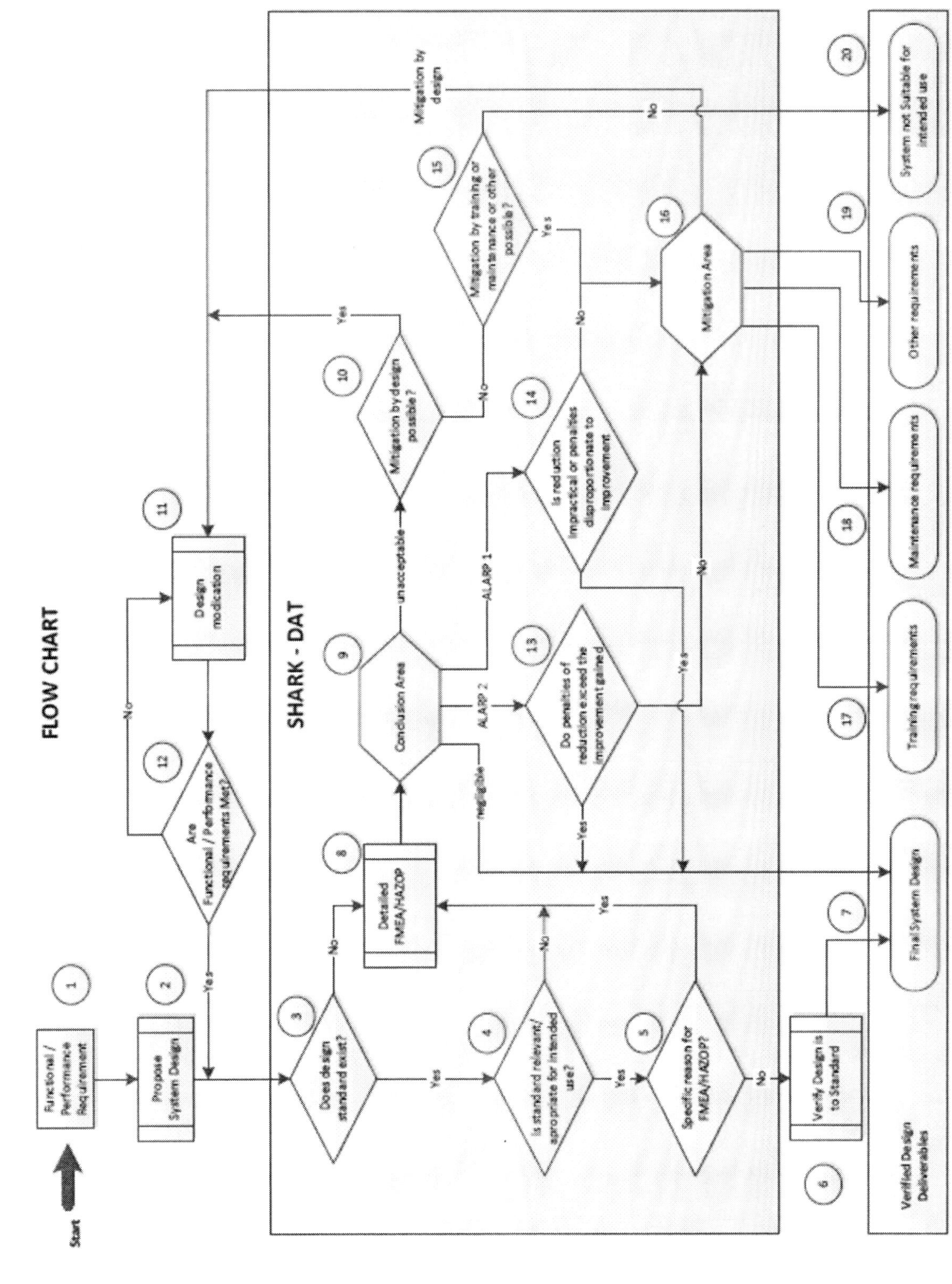

Figure 8: Flowchart

Noise, Vibration, and Repeated Shock Exposure on High Speed Craft: Measurement, Mitigation, Preventative Design.

T Coe and **P Moxley,** Frazer-Nash Consultancy Ltd

SUMMARY

In order to limit exposure to physical agents (noise, vibration, and shock) it is necessary that due consideration to countering the generation of these agents be given at every stage of the lifecycle of a platform, from design through to operation and retrofit activities. This paper summarises a three-phase systems-engineering approach to physical agent control, covering preventative design, measurement, and mitigation. It also details typical activities associated with each of these phases and demonstrates how the phases interact with one another to produce positive design feedback loops for the life of the platform, and 'lessons learned' for use with future platforms.

NOMENCLATURE

SEAT	Seat Effective Amplitude Transmissibility
SED8	Daily equivalent static compression dose
SRS	Shock Response Spectrum
VDV	Vibration Dose Value, $ms^{-1.75}$

1. INTRODUCTION

Surveillance and search and rescue craft need to operate in a "maximum effort" capacity, transiting to incidents quickly to ensure maximum impact. This presents operational challenges, including injurious noise, shock and vibration exposure (for the non-technical reader, shock is defined as a short duration, high magnitude vibration event). Understanding and mitigating noise and vibration requires three key activities: preventative design, measurement and mitigation. This paper discusses challenges around these areas using a whole system approach.

Legislation for noise [1], vibration, and shock [2] shares common themes, requiring that operators demonstrate that levels of exposure are below a specified threshold, and mandating actions for both operators and users in order to mitigate exposure in the event that thresholds are exceeded. The Health and Safety Executive (HSE), as prosecuting organisation in the UK, specify that engineering and organisational solutions to noise, shock, and vibration exposure need to be employed as the primary means of mitigation. Exposure to non-propagating wind noise is a high risk to crews on open high speed craft. The use of a head and torso simulator (HATS) is effective in measuring noise, establishing conditions which produce non-propagating wind noise at significant levels, and assessing performance of comms and hearing protection. These data highlight problems relating to speed, and help generate engineering control actions, as well as organisational and Personal Protective Equipment (PPE) solutions, which are cheaper and easier to implement. It also facilitates rapid solution iteration, enabling users to test PPE or comms solutions in situ.

Shock and vibration measurement on high speed craft has enabled practical issues to be overcome, and severe vibration exposure and events which may cause injury to be identified, and then mitigate them through technology or organisational changes. For noise and vibration, it is essential to consider control at source, especially for new platforms where designs can be easily influenced. HydroDyna modelling enables quantification of the impact of hull features on vertical vibration, and consequently select hullforms to minimise vibration exposure while balancing other requirements. Similarly, noise and airflow modelling allows for the design of wind noise mitigation strategies.

2. SYSTEMS-ENGINEERING APPROACH

This paper presents a systems-engineering approach to dealing with exposure to noise, repeated shock, and vibration on high speed craft. This systems-engineering approach splits activities into three phases: preventative design, measurement, and mitigation. An understanding of how outcomes from each of these phases inform decisions made in other phases is an efficient way to ensure that exposure to physical agents is minimised at every step of the high speed craft design process. A flow diagram illustrating the activities associated with each of the phases in a systems-engineering approach, and how these activities interact, is given in Figure 1.

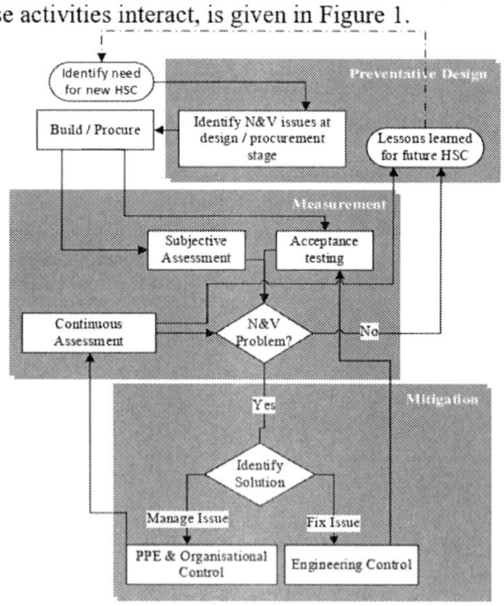

Figure 1 - Flow diagram for systems-engineering approach to noise and vibration control.

The "Preventative Design" portion of the process occurs during the procurement or design phase, and involves both design engineers and subject-matter experts collaborating to ensure best procurement and design practices are utilised to design out noise vibration issues.

"Measurement" can occur at any phase of the process past the build or procurement phase, and acceptance testing is usually the first incidence of this. The measurement phase is used to inform the mitigation phase, but is also expected to provide learning opportunities to be fed forward to the next procurement and design phase.

The "Mitigation" phase is intended to occur after the measurement phase, with technical experts providing engineering control actions, PPE specification, or advice on organisational changes to reduce any observed exposure problems. Mitigation actions usually end in a "loop" using the flow diagram model: Engineering control activities need to be assessed in the measurement phase, and adjusted if found lacking. Where PPE or organisational methods are used, there needs to be a process of continual assessment to ensure that the situation does not worsen, and that control measures are effective. Mitigation stage activities are also expected to inform the preventative design stage, by highlighting what fixes were required and what problems were not addressed in the design stage.

Each of these phases, including examples of each of the activities in this flow diagram, are discussed in greater detail in the following sections.

3. MEASUREMENT

3.1 NOISE

Research undertaken by Frazer-Nash has shown that noise on high speed craft can typically be broken down into three categories: Machinery Noise (main engine, turbo-charger, air intakes etc.), Situational Noise (Comms, sirens and alarms, weaponry etc.) and Environmental Noise.

Machinery noise can be thought of as the noise resulting from any of the mechanical systems required to drive the boat forwards; main engine, air intake, turbocharger and so on. These sources are generally continuous throughout operations, and their impact is typically easy to predict and quantify, as they rely on consistent mechanical systems. The resultant increase in load on the engine and engine / shaft rotation speed caused by increasing boat speed means that machinery noise is typically very dependent on operating speed

Situational Noise is anything which results from a system which provides additional functionality over and above the basic operation of the boat, e.g. sirens on police boats, VHF radio systems, and in-boat comms systems. Situational noise is usually either; used so intermittently that the overall impact on cumulative noise exposure is minimal; so high in magnitude that its use dominates the exposure pattern (sirens, gunnery, etc.), or; is relayed at a volume which can be adjusted to comfortable or tolerable levels by the user (VHF, comms etc.).

Environmental Noise is anything which results from the interaction of the boat with its operating environment, e.g. wind noise, wave noise, spray, non-electronic comms between personnel. Of particular interest to high speed craft is what is termed "non-propagating wind-noise". This is noise produced by the movement of wind past the ear and other protruding surfaces. In lab trials carried out at wind speeds equivalent to over-ground speeds of 43 – 62 knots, [3] shows that noise levels at the ear produced by turbulent flow over the windscreen of a motorcycle helmet can range from 91 to 107 dBA, and this finding has been borne out in field trials carried out by Frazer-Nash. This makes non-propagating wind noise one of the dominant factors in overall noise exposure at high speeds.

As a result of this strong speed dependency, we can think of the overall noise profile on a high speed craft as following the model laid out in Figure 2.

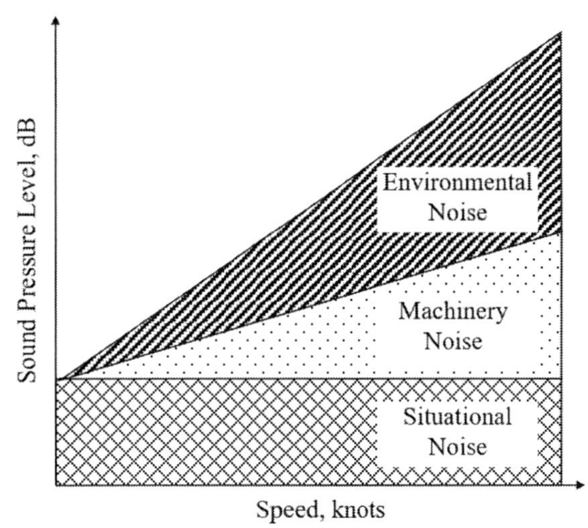

Figure 2: Conceptual model of contribution to overall noise level as a function of vessel speed

As speed increases, the relative contribution of both machine noise and environmental noise increases, with environmental eventually becoming the dominant noise source.

Due to its inability to generate non-propagating wind noise equivalent to the human ear, the ability of a standard instrumentation microphone to measure noise on high speed craft decreases with increasing speed. To bypass this shortcoming, alternative measurement methods are required.

There are two main alternatives to using a standard instrumentation microphone: the Microphone in Real Ear

(MIRE) method, and a Head and Torso Simulator (HATS). The MIRE method involves the insertion of a small microphone into the ear of a user in-situ, whilst the HATS method uses a calibrated model of the human head and torso with microphones located at the approximate location of the eardrum within the head of the dummy. The HATS method was selected as it more readily meets the burden of repeatability, in that it removes the influence of subject physiology and behaviour.

As part of a larger body of work looking at noise in high speed craft in the defence sector, Frazer-Nash undertook a series of measurement trials comparing the results of the shoulder-mounted instrumentation microphones and a HATS on a Rigid Inflatable Boat.

The trial was conducted in March 2017 out of HMNB Devonport. The trial comprised two runs (up-river and down-river) of eight speeds in five-knot intervals from 0 to 35 knots, and a final two runs at max speed (Which was measured at around 38 knots), in each of three occlusion conditions (unoccluded, Gecko helmet, Gecko helmet with visor). A general discussion of the findings of this trial is outside the scope of this paper, and we will instead focus on the comparison between HATS and microphone data. A picture of this experimental setup, with the HATS in the unoccluded condition, is given in Figure 3, and the corresponding results are given in Figure 4.

Figure 3: Trials setup, with standard instrumentation mics on the shoulder of the HATS.

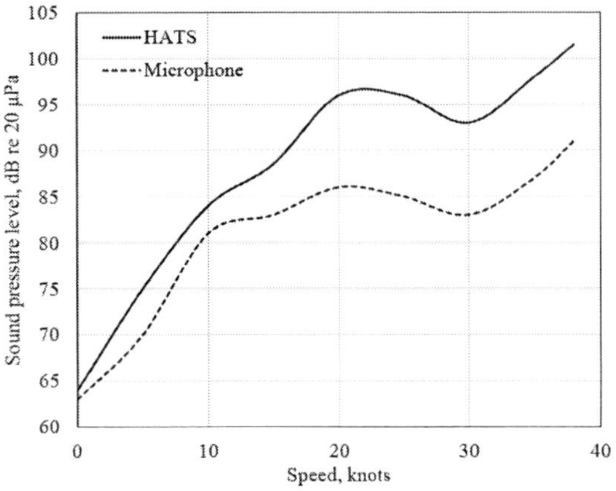

Figure 4: Comparison of HATS and Microphone measurements for the unoccluded condition

Levels varied from around 63 dBA at idle, up to 101 dBA at maximum speed. These correlate well with findings in [4], which logged a range of 85 – 120 dBA for levels external to the ear of a real subject at speeds from 8 to 52 knots. The difference between the two measurement systems is observed to increase with increasing speed, with a maximum under-prediction of 11 dB at 35 knots. This 11 dB difference should be given some context: the time required to reach the upper exposure limit value (the level above which an employer is obligated to halt working activities until noise exposure levels can be reduced) is 5 hours using the microphone-measured sound pressure level, but less than 30 minutes using the HATS measurement. In the scenario where noise exposure is assessed using the microphone measurement, a user could potentially be exposed well above action values, presenting a real risk of injury to personnel.

The conclusion from this measurement series – i.e. the prevalence of non-propagating wind-noise and the dependency of overall noise levels on vessel speed provide clear indication of the mitigation actions that should be taken: organisational changes to review operating speeds, ensuring appropriate hearing protection is available, investigating whether screening can be added to the vessel to reduce the wind speed at the user. It also shows that the use of a standard instrumentation microphone is inadequate for assessing noise exposure on high speed craft.

3.2. SHOCK AND VIBRATION

Over the preceding two decades the understanding surrounding the measurement of shock and vibration exposure in HSC has grown immensely. These advances have shown that, in order to gain a true understanding of the environment that the user is affected by, a number of key considerations are important.

The first of these is bandwidth and filtering. The measurement of whole body vibration is addressed in British and International standards, and deals variously with how to assess affects and the requirements for measurement instrumentation. ISO 8041:2005 in particular provides explicit nominal frequency ranges that a measurement system must conform to, those being 0.5 – 80 Hz. In order to avoid aliasing effects, it is necessary to capture data at a sampling frequency at least double the upper frequency of interest, i.e., at least 160 Hz. In addition to an understanding of sampling requirements it is also necessary that whoever performs the assessment have a good grasp of signal processing and analysis, which allows them to identify specific features of the vibration environment which may contribute to the vibration exposure. Vibration measurement on high speed craft also requires that the analyst be familiar with the specific features of a high-shock environment.

The second consideration is the way in which the *exposure* of the vibration is assessed and quantified, which is distinct from how the vibration is measured in objective terms. In the UK, the preferred method of vibration exposure assessment is the vibration dose value (VDV). This is a dose relationship, which may be described mathematically as '*the integral of the fourth power of the weighted instantaneous vibration in* $ms^{-1.75}$'. It relates the magnitude of the vibration and the time for which the vibration is experienced to a single figure value. VDV is preferred to other common metrics, such as root-mean-square (r.m.s) 8-hr equivalent acceleration (A(8)) because it is sensitive to signals which have a high crest factor (i.e. a signal which contains large peak or shock events which are many times greater than the r.m.s acceleration level). VDV has also been demonstrated to correlate well to metrics preferred in other countries for both continuous vibration and vibration containing shocks.

A comparison of VDV and the Shock Response Spectrum (a response derived from a mathematical model of the human body) in response to a shock waveform is given in Figure 4, where it can be seen that the relationship between the two metrics is highly linear across a large range of accelerations.

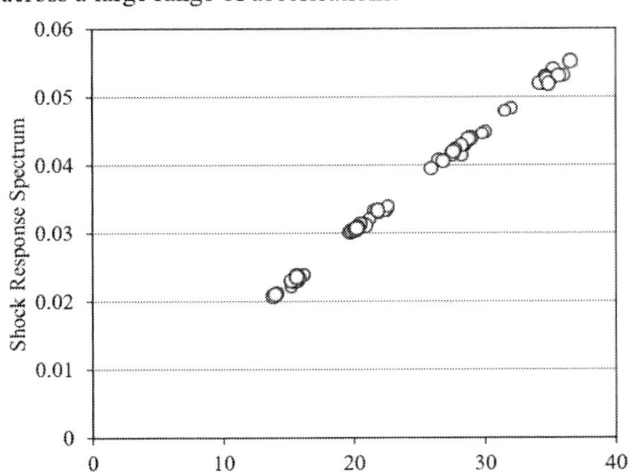

Figure 4: Comparison of SRS and VDV exposure values for single impacts.

A comparison between seat effective amplitude transmissibility (SEAT) values derived using the VDV method and the Daily Equivalent Static Compressive Dose Value (SED8) is shown in Figure 5. Whilst this data is not directly comparable as a method of exposure assessment (as it is being used to determine the amount of energy passed from the deck of a boat to the surface of a shock mitigating seat), it does show that the two metrics correlate well over a range of vibration magnitudes.

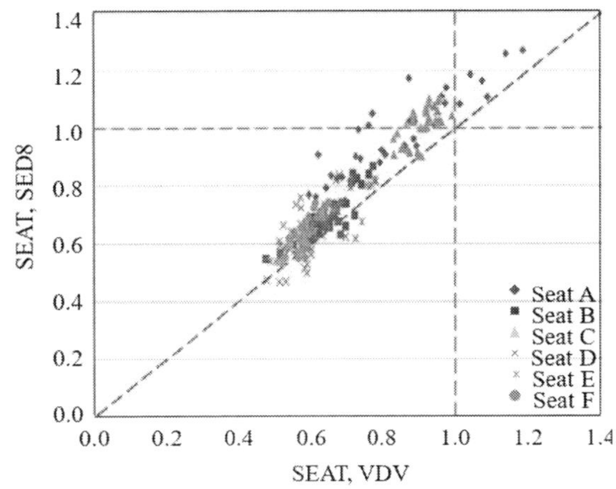

Figure 5: Comparison of SRS and VDV exposure values for single impacts.

British and international standards for vibration exposure measurements highlight the distinction in sensitivity between input locations, stipulating the use of inter-axis scaling factors and changes in frequency weightings, specifically for input at the backrest, seat and feet. Whilst these allowances enable an analyst to quantify vibration input to the body in a standardised way, it fails to factor in the pronounced effect that posture can have on the human response. Factors such as the "uprightness" of the upper body can change resonance frequencies in the vertical axis [5], and how the feet are supported relative to the motion of the environment changes both the sensitivity of the person experiencing the motion, and where in the body they experience said motion [6]. Additionally, one can also imagine the impact that loads worn on the back are likely to have on the compressive forces of the lower spine and on the magnitude of forces at the upper spine as the body gives in the fore-and-aft or lateral axes. Consideration, therefore, needs to be given to the way in which occupants are seated and how they are loaded when considering vibration assessment. Standards, including methods for assessing seat performance, also neglect the impact of the 'active control' offered by the occupants own musculoskeletal system (i.e. the subject's ability to mitigate some of the effects of their motion relative to the boat by bracing or by exerting themselves in the opposite direction to large shocks to reduce body motion). These two factors present a certain level of uncertainty with regards to understanding vibration measurements: on the one hand, additional load and suboptimal posture may produce a greater response to vibration which is not captured by the measurement process, whilst on the other, the subject's own ability to help mitigate impact may mitigate the response, which again is not captured by the measurement process. Consequently, any person undertaking vibration measurements needs to also observe the working environment and occupant use patterns to provide context to the single figure values which are usually the main output of vibration exposure assessments.

There are other, more practical considerations to take into account when measuring vibration: The nature of environments with a high shock component is such that people may commonly lose contact with the seat measurement surface, which produces artefacts in the time history which are not truly representative of the vibration experienced. This can be mitigated by the use of a pressure switch which prevents vibration being measured whilst the occupant is not in contact with the seat, or sections of time history can be removed based on observation or recordings of occupant behaviour taken with a video camera; The method used to secure an accelerometer to the deck (i.e. cyanoacrylate, beeswax, adhesive tape, magnetic bases) is likely to have an impact on the resonance frequency of the measurement system and can distort measurements at high frequencies, and particular care needs to be taken in selecting the most secure and appropriate method, such that the accelerometers do not come lose or become damaged during the trial – insufficiently robust mounting of accelerometers results in lost data, delays to schedule (especially in the case of acceptance testing), and increased costs associated with the overall project.

4. MITIGATION

4.1 NOISE

Now that we have understood the measurement approach for noise on high speed craft, we have some idea as to what the most efficient and effective way to mitigate noise exposure problems. We can broadly think of the control actions as falling into three categories, as detailed in Figure 1: Engineering Control, Organisational Control, and PPE.

4.1 (a) Engineering control

Engineering noise control activities typically centre on remedial work to machinery, increasing acoustic absorption in machinery spaces, or sourcing lower-noise alternatives to existing equipment.

The addition of acoustic absorption in machinery spaces is likely to be expensive, only applicable in larger craft with a dedicated machinery space, and to only give a relatively small return on investment. On open boats, particularly those with an outboard propulsion system, there is no practical means of adding additional absorption. It is also broadly accepted that the addition of extra insulation is a poor way to use a boat's weight budget.

Remedial work on existing machinery is likely to only be applicable where there are demonstrable flaws in the machinery which should be caught as part of general maintenance on the platform, e.g. worn bearings, damage to ventilation grills, wear and tear on any acoustic enclosures. Sourcing of low-noise replacements is likely to have the greatest impact on noise levels in high speed craft, but it should be acknowledged that there is a substantial increase in cost and in vessel uptime resulting from time taken to source and retrofit appropriate parts.

The most effective way to limit retrofit costs and platform down time is by using a whole-system approach to noise control, beginning with sourcing low-noise components or platforms at the design/procurement stage, and including a measurement phase which identifies problem components.

4.1 (b) Organisational controls

Where engineering controls have been mooted (either as being impractical or providing insufficient benefit), it is possible to manage noise exposure using organisational means. For example, where a platform has been identified as having a noise problem, it would be beneficial to address how the platform is operated. It may be useful to consider addressing the following questions: Could the same goals be achieved whilst operating at lower speeds; could the same goals be achieved with fewer personnel; Are there ways to form a rota such that noise exposure burden is evenly split between all personnel; Could personnel be relocated so that they are behind windshields, or away from air intakes etc.? This is obviously not an exhaustive list, but questions along these lines are likely to generate cost-effective and easy to implement fixes to some aspects of the noise exposure problem.

Another aspect to consider would be dosimetry, or regular hearing checks. A dosimetry system (of which a great many off-the-shelf solutions exist) would provide users and line-management a real-time understanding of how specific users are being exposed to noise and allow for staffing and rota decisions to be made which limit individual exposure

4.1 (c) Personal protective equipment

The view of the Health and Safety Executive (HSE) is that PPE should be used as a last line of defence, having exhausted all other remedial control measures, i.e. it is not viewed as acceptable to ignore, for example, a loud turbocharger in favour of providing better hearing protection. Instead, any remedial work which is reasonably practicable should be implemented, and hearing protection used to reduce exposure further if no other gains can be achieved through engineering methods.

In order to specify appropriate hearing protection, it is necessary to take inputs from the measurement process to establish what the cause of the noise exposure is, and to engage with end users and stakeholders to understand what other requirements need to be met, for example: Would the hearing protection interfere with life vests or helmets?; Is there a requirement to keep user payloads as low as possible?; What is the pattern of the noise which is being controlled by hearing protection – is it intermittent or continuous, does it have a strong impulsive component?; Is there a need to combine hearing protection and a comms system? Each of these considerations – along with many other more usual factors, such as cost and form factor – dictates the characteristics of an ideal hearing protection solution.

One of the complications associated with hearing protection is that users of high speed craft are often required to maintain communications in order to provide feedback as to mission objectives. Both high noise levels and effective hearing protection will degrade verbal communication, and so a common tactic to combat this is to include a comms system in the hearing protection solution, which introduces further variables to the noise exposure profile. Work undertaken by Frazer-Nash in support of noise management within the MoD included the assessment of a hearing protection solution with built in comms and a "super-hearing" function (essentially a pass-through from external microphones which amplifies external sound for increased situational awareness) at a range of speeds. Both the comms system and the super-hearing function, with levels optimised by an experienced technician prior to testing, were shown to significantly increase user noise exposure over the protected exposure level (for the comms exposure case) and over the un-protected level (for the super-hearing case). A systems-engineering approach which involves appropriate measurement and stakeholder engagement is the most efficient way to ensure adequate PPE is procured, and that the burden placed on PPE is reduced through preventative design and engineering control methods.

4.2 SHOCK AND VIBRATION

The same categories of mitigation and control can be thought of as with noise. However the background understanding is perhaps less well developed. Here we will consider all three areas from the perspective of shock and vibration measured at the deck.

4.2 (a) Engineering control

Engineering control for shock mitigation primarily involves the retrofitting of additional equipment, the assumption being that *post hoc* adaptation of hull form is so far beyond onerous and impractical to even consider.
A variety of control measures are feasible in this regard;
Seats and cushions: seats which have been designed so that they are able to move in response to shock and vibration stimuli, or cushions with compressible foam or air bladders;
Consoles: Similar in many ways to shock mitigating seats, but typically a single unit on which several seats are mounted;
Deck coverings: usually comprising some sort of relatively thin material which can deform in response to shock and mitigate some of its impact on standing users
As with noise, an understanding of the outputs of the measurement phase is key to selecting the correct engineering control solution, as is an understanding of the operating parameters, weight restrictions etc. For example, it may be possible to achieve a significant reduction in the time taken to reach exposure limit values by installing a shock mitigating seat with a vertical travel distance of over a meter, but the weight of such a system, and it's suitability for a marine environment would usually preclude it, meaning that it would be necessary to select something with a smaller impact which does not impact performance, such a shock mitigating cushion. Or, it may be that installing shock mitigating seats is practical but, due to the boats hullform, the lateral component of the shock is actually the most injurious, and so consideration needs to be given to the seats form, for example ensuring that the seat back is contoured, or has arm rests to reduce lateral impact.

In recent years the focus in shock mitigation has been on seating, as this is where the most significant impacts can be made. However, the impact of cushions and deck coverings – which are known to have a typically smaller impact than seats - should not be overlooked, as a small reduction in exposure level (as measured by VDV or similar) can have a large impact in terms of the time taken to reach the exposure action and limit values

As with noise, the HSE mandate that vibration exposure be reduced to levels 'As Low as Reasonably Practicable' (ALARP), and are fully cognisant of the limitations of the 'Practicable' aspect of this. Exposure to vibration is a very real fact of operating HSC, and there is no expectation that levels be completely eliminated, or that the operation of a vessel be so hamstrung by engineering control that it cannot operate safely and effectively. Any engineering solution which is applied needs to incorporate a good understanding of the environment in which the boat operates and the operating requirements of the platform under assessment.

4.2 (b) Organisational control

Organisational control involves changing working practices to control exposure. This can involve the following activities: Management of personnel to reduce repeated exposure, for example rotating crews; Training of crews to minimise impact through improved driving and coaching [7]. There is relatively little formal literature on this approach but there are undoubtedly 'good' and 'bad' coxswains and both training and coaching could be used to improve performance in a learnt skill; Training to minimise risk from impacts for passengers and crew alike. For example a programme in the USA [8] has investigated targeted strength and conditioning programmes for US military personnel. Similarly the Royal Navy has developed strength and conditioning guidance for boat operators [9]; Operational planning: with evidence that fatigue and long exposure are factors in shock and vibration injury and discomfort [10], treating boat crew more like air crew with mandated breaks and flying limits may be appropriate. The Federal Aviation Authority in the USA limits pilots to 9 consecutive hours flying and duty periods of 14 hours interspersed with 10 hour rest periods. Military operational planning and SAR operations might preclude these rules but comparing this to the situations described by Dobbins et al. [11] illustrates the issue.

The Health and Safety Executive mandate Health monitoring/surveillance of personnel exposed above the exposure action value and there are a number of commercial providers of systems. Two key confounding issues are:

Low completion rates: To gather data immediately following sea transits requires paperwork or IT access and users are likely to be cold, wet, tired and consider health surveillance to be a low priority;

Organisational culture: One can envisage a situation where health issues from boat operations are under reported either because the patient fears the consequences or it does not meet the 'macho' culture of the organisation.

In both these instances lessons can be learnt from aviation where pilot logs are routinely filled in and safety culture is ingrained.

4.2 (c) Personal protective equipment

PPE is defined here as equipment which is either body worn or issued to personnel rather than installed on the craft.

This is an area where the industry has yet to develop with respect to WBV. There are however a couple of products on the market, taking the form of exoskeletons which aim to reduce the use of leg muscles for shock mitigation [12] and hence improve protection. The market has not seen wide adoption of such devices but published data from the manufacturers shows an effect.

Personal dosimeters, which aim to measure operator's exposure, are not well developed, standardised, nor licenced. As with other forms of PPE, the difficulties in quantifying an actual reduction in risk of harm contribute to the difficulty in the maturing of this market.

5. PREVENTATIVE DESIGN

The aim of preventative design is to ensure that there are no noise, shock, or vibration exposure issues for end users, prior to ever stepping foot on the boat. In order to do this, there are two activities which should be considered: Firstly, to engage high speed craft designers and users in order to ensure that, where off-the-shelf solutions are being sourced for propulsion etc., low-noise items are selected; Secondly, that best design practices are followed to reduce noise, shock, and vibration.

The reader is directed to [13] for a robust guide to the inclusion of human factors principles at the design stage.

5.1 NOISE

From our earlier discussion of mitigation techniques, we know that there are a variety of actions which could be implemented in order to reduce noise exposure at the design stage: Sourcing low-noise propulsion equipment; designing the high speed craft such that personnel are seated away from any sources of noise; locating personnel behind screens which reduce the velocity of wind flowing past them.

The most direct way of ensuring low noise components is to engage subject matter experts with proven experience in procurement for high speed craft. They will be able to perform market surveys and engage with the original equipment manufacturer (OEM) in discussions on noise. Further to this, they may be able to advise what the "cutting edge" solution looks like, and give advice as to whether alternative propulsion systems could be used, such as hybrid or electric systems. They should ideally form a part of the design team, in order to ensure that the end solution meets all system requirements.

Designing high speed craft to reduce wind speed at the user is largely a case of installing shielding at strategic locations to deflect wind away from the user, or of locating the user in a recessed or enclosed space away from the wind. Whilst the latter of these approaches may be feasible in larger high speed craft, they are not practical for open boats such as ribs and dinghies, in which case it may be feasible to install wind shields. The efficacy of a wind shield depends on its form factor and its location, and will likely interact with limitations placed on the overall drag of the vessel. Placing a large windscreen at the front of the vessel is likely to reduce wind speed for all users compared to the fully open condition, but will have serious implications for drag of the vessel, whereas small individual windshields are likely to have a smaller impact on drag, but may be less effective at reducing wind speed – you may also be limited by safety factors and the ability to install individual windshields.

5.2 SHOCK AND VIBRATION

Preventative design in the field of shock and vibration mitigation is deeply entwined with seakeeping design and the development of seakindly hull forms. By addressing the issue during the design phase and considering the entire craft system it is possible to maximise any performance gains.

An issue for designers, builders and procurers of HSC, in particular military HSC, is the constant need to balance capability requirements with the limitations of the boat. This can drive seakeeping aspects of craft performance down the collective priority list.

However there are numerous potential benefits to considering this differently. Considering shock and vibration mitigation as part of a whole system approach to design [13] can lead to a better performing craft overall.

Knowledge of the environment on board gained from measurement of similar craft or from simulation or model testing is important for quantifying the benefits. Considering solutions at the design phase allows major features of the design to be optimised with human performance on board in mind.

In order to reduce peak accelerations and jerk and increase the duration of impacts it is necessary to reduce pressures on the craft from slamming. This has the effect of increasing displacements and hence likelihood of shipping green water.

Features such as the following should be considered:

- **L/B ratios**: Higher L/B ratios give a finer craft with lower peak accelerations.
- **Shearlines**: A shearline that sweeps upward in the bow region provides additional freeboard, volume and reduces the chance of shipping green water.
- **Deadrise angles**: Higher deadrise angles reduce impact pressures while increasing power requirements.
- **Flare**: Defined as the derivative of local deadrise, can be tuned to 'stiffen' pitch motions.
- **Tube profiles**: Tubes and foam pseudo-tubes can prevent bow 'stuffing' and reduce peak accelerations with shaping and correct pressure [14].
- **Spray rails**: Steps and chines have all been cited as having positive seakeeping benefits if correctly designed [15, 16].

It is however often the case that these features are 'expensive' in design terms. A number of these are illustrated in Figure 6 comparing a military craft designed with apparently low signature as a design driver and an all-weather lifeboat. The apparent desire to maintain a low profile in the military craft drives down the wheelhouse height and flattens the shearline in the bow. The bow shape of the military craft appears to include a constant deadrise angle compared to the twist of the lifeboat, with an increasingly fine angle towards the bow giving reduced peak accelerations. Overall the constraints of interoperability, signature and aesthetics within the military context override seakeeping features seen in the RNLI vessel.

What must be considered though is that shock and vibration exposure is intimately linked to the overall capability of the craft in real conditions and therefore improvements here can improve actual useable capability.

Rapid, medium fidelity simulation such as the Explicit Dynamics Finite Element Code HydroDyna is a useful comparative tool for assessing the benefits of different hull form features.

Figure 6: (Top) US SOCOM SeaLion II Special Operations Stealth Boat; (Bottom) Shannon class lifeboat, RNLI

Determining the benefit of these features is key to understanding the cost benefit analysis in design terms. Combining simulation, knowledge of mitigation systems and an understanding of human response to motions allows a human centred operational envelope for a craft to be developed and used as key decision making tools for selecting designs and making ALARP arguments.

6. ACKNOWLEDGEMENTS

We would like to acknowledge DE&S Commercially Supported Shipping for allowing the use of data.

7. REFERENCES

1. HEALTH AND SAFETY EXECUTIVE, '*The Control of Noise at Work Regulations*', 2005
2. HEALTH AND SAFETY EXECUTIVE, '*The Control of Vibration at Work Regulations*', 2005
3. LOWER, M.C., HURST, DW, THOMAS, A, 'Noise Levels and Noise Reduction under Motorcycle Helmets', *Inter-Noise*, 1996.
4. SEIDMAN, MD, WERTZ, AG, SMITH, MM, JACOB, S., AHSAN, S.F., *'Evaluation of Noise Exposure Secondary to Wind Noise in Cyclists"*, Otolaryngology – Head and Neck Surgery, Vol 157 Issue 5, 2017
5. KITAZAKI, S, GRIFFIN, MJ, *'Resonance behaviour of the seated human body and effects of posture'*, Journal of Biomechanics 31, pp 143-149, 1998

6. JANG, H, GRIFFIN, MJ *'Effect of phase, frequency, magnitude and posture on discomfort associated with differential vertical vibration at the seat and feet'*, Journal of Sound and Vibration, 229(2), 2000
7. PIKE, D, '*'The Complete RIB Manual: The definitive guide to design, handling and maintenance'*, Adlard Coles Nautical, London, UK, 2013
8. DOBBINS T, ROWLEY I, CAMPBELL R, *'High Speed Craft Human Factors Engineering Design Guide'*, Human Sciences & Engineering Ltd, 2008.
9. FINNEMORE, R, *'MoD Risk Reduction Measures to Counter the Effect from Maritime Whole Body Vibration'*, presented at the 51st United Kingdom Conference on Human Response to Vibration, Institute of Naval Medicine, Gosport, 2016
10. COLWELL J, GANNONG L, GUNSTON T, LANGLOIS R, RILEY M R, COATS T, *'Shock Mitigation Seat Test and Evaluation*, Royal Institute of Naval Architects, 2011
11. DOBBINS T, HILL J, MYERS, S, *'Fatigue in Military Operations: High Speed Craft Repeated Shock and Other Factors,* presented at the 51st United Kingdom Conference on Human Response to Vibration, Institute of Naval Medicine, Gosport, 2016
12. BRIDGER R, ASHFORD, A, WATTIE S, FISHER I, PISULA P, *'Pilot Study of an Exoskeleton for the Lower limbs'*, presented at the 51st United Kingdom Conference on Human Response to Vibration, Institute of Naval Medicine, Gosport, 2016
13. SELL TC, ABT JP, CRAWFORD K, et al. *'Warrior Model for Human Performance and Injury Prevention: Eagle Tactical Athlete Programme (ETAP)'*, Journal of Special Operations Medicine Vol 10, Ed 4/ 2010
14. TOWNSEND N, WILSON PA and AUSTEN S, *'What influences rigid inflatable boat motions?'* Proceedings of the Institution of Mechanical Engineers, Part M: Journal of Engineering for the Maritime Environment, Volume: 222 issue: 4, page(s): 207-217 2008.
15. http://www.navatekboatbuilders.com/sb-40.html
16. CAMPBELL L. *'Notes on stepped hulls'* presented at High Speed Boat Operators Forum, Gothenburg, Sweden, 2012

8. AUTHORS BIOGRAPHY

Tom Coe is a Senior Consultant with Frazer-Nash Consultancy currently on secondment to the Naval Design Partnering as the High Speed Craft Group Manager and Technical Lead of the MoD's Maritime Whole Body Vibration Project. He has an extensive background in WBV in small fast craft having completed a PhD in the subject from Southampton University, conducted numerous at sea and laboratory trials and is a member of a number of international committees on the subject.

Phill Moxley is a Senior Engineer working in the noise and vibration group of Frazer-Nash Consultancy, where he has been involved with the measurement and assessment of noise and vibration on a variety of marine platforms. He is currently on part-time secondment to the Naval Design Partnering as a whole body vibration modelling expert. Prior to this he undertook a PhD in human responses to vibration at the Human Factors Research Unit in the Institute of Sound and Vibration Research, University of Southampton.

FULL-SCALE SEAKEEPING TRIALS OF AN ALL-WEATHER LIFEBOAT

F Prini, S Benson, R W Birmingham and **R S Dow,** Marine Offshore and Subsea Technology Group, School of Engineering, Newcastle University, UK
L J Ferguson, P J Sheppard and **H J Phillips,** Royal National Lifeboat Institution, UK
M C Johnson, J Mediavilla Varas† and **S Hirdaris*,** Lloyd's Register, UK

(†work initiated whilst at Lloyd's Register. Now at Materials Innovation Institute, NL)
(*work initiated whilst at Lloyd's Register. Now Visiting Research Scholar at Maritime Technology Group, Department of Mechanical Engineering, Aalto University, FI)

SUMMARY

The Severn class of the Royal National Lifeboat Institution (RNLI) has recently undergone extensive sea trials as part of collaborative research with Newcastle University and Lloyd's Register. The scope of the trials was to investigate the seakeeping behaviour of the craft in real operational conditions. A Severn lifeboat was fitted with a number of sensors to measure hull girder loads and local loads due to wave and slamming. Time-domain and spectral analyses were applied to the data collected and the first results are presented. The Response Amplitude Operators (RAOs) of vertical bending moment are also compared against numerical predictions. As part of a major plan that combines numerical methods and experiments to investigate the sea loads sustained by the structure during operation, the sea trials contribute to validating the wave loads predicted numerically and allow to account for slamming-induced load effects.

NOMENCLATURE

Δ	load displacement, in kg
GPS	global positioning system
$H_{1/3}$	significant wave height, in m
kg	kilogrammes
kn	knots
λ	wavelength, in m
LCG	longitudinal centre of gravity, in m
LOA	length overall, in m
LWL	length waterline, in m
m	metres
RAO	response amplitude operator
RPM	revolutions per minute
T	draught, in m
V	speed, in kn
VBM	vertical bending moment, in Nm

1. INTRODUCTION

The RNLI is a charitable organisation that exists to save lives at sea around the coasts of the British Isles, covering all political jurisdictions i.e. the United Kingdom, the Republic of Ireland, Isle of Man and the Channel islands. Founded in 1824 it is independent of any Government and provides over 85% of all waterborne Search and Rescue (SAR) assets to these governments, free of charge to both government and end-user. The RNLI operates a fleet of strategically placed all-weather lifeboats which provide cover out to 100 miles from the coast in all weathers, and a tactically placed fleet of Inshore Lifeboats which ensure that 90% of casualties within 10 miles of the coast can be reached within 30 minutes from notification.

The first All Weather Severn class lifeboat entered service in 1995, having been designed with a 25 year operational life. In common with the majority of the RNLI's fleet, of which the Severn is the largest, they are constructed using advanced composite materials. With an outstanding service record and well-liked by the crews the early boats were reaching the end of their original design life and would require a program of either replacement or life extension.

Early work showed the hull panels were capable of an 80 year working life and an engine replacement project was already underway replacing the original engines for lighter more powerful units.

The trials reported in this paper form part of a major research project, undertaken in collaboration with Newcastle University and Lloyd's Register, to consider the approach to structural design that should be adopted by the RNLI. While these trials will help validate the decision to extend the operational life of the Severn class, the complete project, which encompasses all the design criteria used by the RNLI, will enhance this and future design projects.

The Severn class lifeboat is shown in Figure 1 and its main particulars are listed in Table 1.

2. DESCRIPTION OF THE TRIALS

The tests took place in the North Sea and consisted of 11 trials conducted at speeds ranging from 5 to 25 knots and in different sea states, with a significant wave height from 0.3 to 4.6 metres. The speed over ground was kept constant throughout each trial by adjusting the engines RPM as required. The trim tabs were set to 'rest' position, fully lifted, and never adjusted.

During each trial, data was collected continuously from on-board sensors. A record sheet was used to for additional information: sea state, wind speed, wind

Figure 1: Severn class lifeboat.

Length overall	17.00	m
Length waterline	15.50	m
Beam overall	5.62	m
Draught	1.37	m
Displacement (full load)	42000	kg
Speed	25	kn

Table 1: Severn class main particulars.

direction, compass heading, turning times, GPS coordinates at the turns, engines RPM, running trim and speed through water.

All the trials were conducted with full fuel tanks at departure. Displacement and centre of gravity position of the lifeboat were already known from previous inclining tests. In addition, draft mark readings were taken in conjunction with measurements of the water density before and after each trial for further estimates of the loading condition.

3. INSTRUMENTATION

An operational lifeboat can be a hostile environment for sensitive electronic equipment to operate in. Based on past experience, the data gathering system was required to withstand the following operating conditions:

- Temperatures range –10˚C to +55˚C
- Accelerations up to +/–20g
- Shock loads up to +/–40g
- Vibrations up to +/–3g continuous

A number of different technologies exist that can be used to gather strain data and these were considered before selecting traditional foil backed strain gauges due to their low cost, high reliability, ease of installation and ready availability. Embedded Fibre Bragg Gratings (FBG) are also considered reliable for measuring strain in composite materials. However the embedded nature of the fibres meant that retrofitting this type of system to an operational lifeboat was somewhat impractical and costly. Additionally, if a FBG fails, all signals "downstream" would be lost. The foil backed strain gauges were bonded to the inside skin of the hull using an epoxy adhesive (see Figure 2).

The range of samples per second (Sps), typically measured in terms of sampling frequency in hertz (Hz), available for strain measurement is large. For example, Vishay Precision Group offers a system used for dynamic signals with scanning rates up to 10,000Hz and up to 1200 channels [1].

Faltinsen [2] conducted experiments on a catamaran equipped with strain gauges and measured multiple structural strain oscillations occurring within a time span of 0.05 seconds. This work highlights the high speed at which structural strain events can happen and emphasises the need for a high number of measurements per second. Should a low sampling frequency be used, then there would be a real possibility that structurally significant strain events would not been recorded. Moreover, since strain can occur also as a result of thermal stress in the material, two thermocouples were placed in areas of elevated temperature: the engine room and the survivor space. The temperature data was used to compensate the strain readings for temperature effects.

Initially it seemed reasonable to measure whole boat accelerations in at least three locations: the forward bulkhead, amidships and the aft bulkhead. However, when evaluating the need for 3 accelerometers it was realised that the numerical results would be based around the LCG and therefore, a single triaxial accelerometer was placed as close to the LCG as possible. Similarly, the rate gyro was placed at the LCG allowing the categorisation of accelerations when post-processing the results.

An understanding of the nature of the ship responses to be measured allowed an informed decision regarding the sampling frequency to be made, taking into account the large amount of data that a high sampling frequency would produce. The resultant measurement types and sampling rates are detailed in Table 2.

Measure	Sensor	Sampling Frequency (Hz)	No. of axes	Quantity (Channels)
Hull girder bending moment	Strain gauge	256	1	18 (18)
Bottom & bow flare slamming	Strain gauge	2048	1	32 (32)
Wet deck slamming	Strain gauge	2048	1	8 (8)
Temperature compensation	Thermocouple	256	1	2 (2)
Whole boat accelerations	Accelerometer	2048	3	1 (3)
Rate of turn	Rate gyro	256	3	1 (3)
Total				**62 (66)**

Table 2: Measurement type and sampling frequencies.

Wiring for all the instrumentation was shielded to prevent electrical interference and led back to a Vishay System 7000 data logger [3] secured on a survivor's seat. The Vishay equipment was successfully used on other RNLI trials and was evaluated and tested again prior to confirming its suitability for this task. The cable routing meant passing cables through bulkheads where it was critical to maintain watertight integrity for survival reasons, should the vessel become damaged, and self-righting ability in case of capsize. Figure 3 shows the cables passing through a gland in a watertight bulkhead and Figure 4 shows the sensor layout.

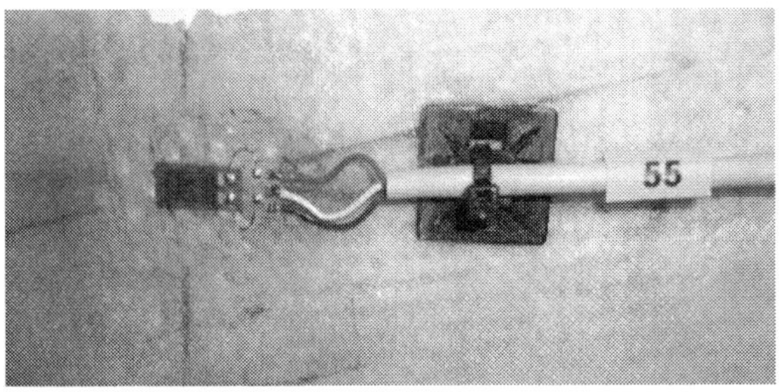

Figure 2: Strain gauge bonded to the hull.

Figure 3: Cables passing through a watertight bulkhead.

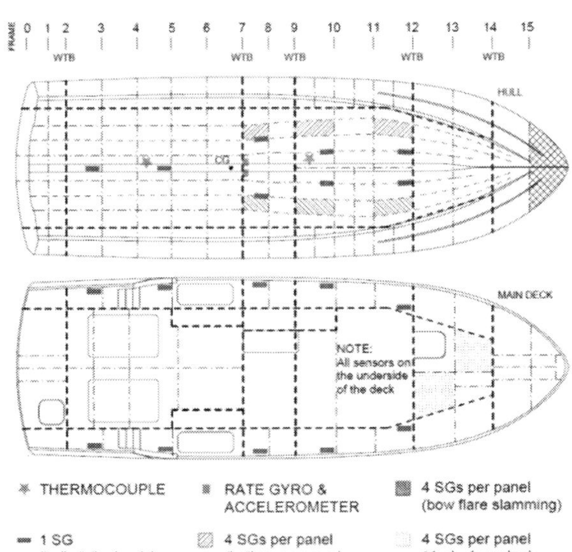

Figure 4: Sensor layout.

4. SEA STATE MEASUREMENT

A measure of the sea state is necessary to correlate motions and structural response to the wave environment. Visual estimates of the wave height and period were the only option before the advent of technological devices for wave measurement. Relationships between observed and measured values were later proposed, as summarised by Lloyd [4]. In addition to the obvious limits to the reliability of visual estimates, even observations made by experienced sailors lack a complete description of the spectral shape. Wave estimation through ship motion analysis has also been attempted [5–7].

Nowadays a more rigorous method is considered a necessity for seakeeping trials. A comprehensive review of techniques for ocean wave measurement is given by Johnson [8]. An important aspect that drives the analysis of data from sea trials is whether the measure is taken at the vessel position or in the trial area. In the first case a particular event in the vessel response can be related to the wave that has generated it. Wave measurements at the vessel can be obtained through on-board systems, such as bow-mounted radars and wave radars. Considerations on the wave data that these technologies provide are not discussed since neither of these options was practical for the following considerations: bow-mounted radars could not be mounted on an operational lifeboat because of the risk of damaging the equipment during rescues; wave radars still come at a prohibitive cost and are not proven to be reliable in rough and confused seas.

Wave measuring devices not mounted on board exclude the possibility to directly relate a ship response to a particular wave since the device will be at some distance from the vessel or will provide averaged values for an area of sea surface. Within this category of devices there are wave buoys, satellite altimetry radars [8] and mathematical hindcast models. As concluded by Johnson [8], wave buoys were considered to be the best compromise combining accuracy, ease of handling, reliability and cost.

A wave buoy, deployed central to the trial area and left free to float for the whole duration of each trial, was used as the principal source of wave information. The buoy was a Directional Waverider DWR-G4 [9–11], which consists of a sphere with a diameter of 0.4m and a weight of 17kg. It was easy to store on deck and to deploy and recover in any sea condition without davits. Because of its small size a danbuoy was loosely tied to it to increase its visibility (Figure 5).

Forward (North), transverse (West) and vertical (heave) displacements were generated through a GPS-based motion sensor. The GPS technology supersedes the need for calibration of accelerometer and compass based sensors and it is not affected by spinning or manual

Figure 5: Datawell DWR-G4 buoy used for the trials. The buoy was left free to float, tied to a danbuoy to increase its visibility.

handling. One potential disadvantage is that wave wash on the GPS antenna or extreme tilting could mask the signal and interrupt the measurement. In fact this only happened sporadically and for a few seconds during the whole deployment time. Displacements are sampled at 1.28Hz, which allows the capture of wave periods as short as 1.6s (≈4m wavelength). In spite of the relative small size of the vessel this was enough to calculate, through a spectral analysis, the response amplitude operators of the ship responses for the whole range of wavelengths of interest.

Additional wave data was also obtained from two wave buoys moored approximately 10 nautical miles N and 23 nautical miles ESE from Tynemouth, operated by the Channel Coastal Observatory (CCO) and the Centre for Environment, Fisheries and Aquaculture Science (Cefas) in the UK. Visual observations were taken at the start and end of each trial. Wind to sea state correlation pictures [12,13] were also used, although the assessment mainly relied on the crew's experience.

5. TRIALS AREA AND TRAJECTORY

The tests took place in the North Sea offshore from Tynemouth (UK). A minimum distance from the shore had to be maintained to avoid systematic errors in the results due to shallow water and coastal effects. Based on theoretical knowledge and experience of the local crews, a minimum water depth of 50 metres, corresponding to a distance of 5-6 nautical miles from the shoreline, was considered appropriate. The 50-metre bathymetry line is also easily identifiable on the charts, which is practical for positioning and orienting the trial trajectory according to the principal wave direction on the day of the trial.

Dedicated seakeeping tests often involve runs at different headings to understand the vessel behaviour in a range of operating conditions. An appropriate sequence can minimise the gap between the runs and the average distance from the wave buoy. The effect of factors that, by changing during the trial, influence the results can also be minimised. Trajectories commonly adopted are described in detail by Johnson [8], who also suggests eight legs at 45° intervals as a suitable option to fully account for the effect of the seaway on the ship responses. An octagonal trajectory represents the easiest option, with the drawback that it can have a large diameter.

The trajectory adopted for the trials was a 'star pattern' devised to test headings from head to following seas, both port and starboard, at 45° increments. This results in the 8-legs star shown in Figure 6, which has the advantage of minimising the average distance from the wave buoy. As commented by Johnson [8], the disadvantage would be the increased time spent in turns, which however becomes irrelevant for a small craft.

The time spent on any one heading determines the number of motion or load cycles recorded and this should be high to create a statistically significant sample. The presence of tidal currents that affect the wave spectra also plays a role. As suggested by Johnson [8] short trials tend to be bound to a larger systematic error than long trials, which are more likely to include slack water and current in both directions. On the other hand the data analysis is simpler if it can be considered that the sea has remained uniform and statistically stationary during the whole trial. This is the case for a limited amount of time, typically 1 to 4 hours [14].

A compromise was found by setting the time spent at each heading so that a minimum of 100 wave encounters, hence motion and load cycles, would have occurred. This was achieved starting from an analysis of wave scatter diagrams for the North Sea [15–17]. The significant wave height was divided into 4 tiers: calm (Douglas 1-3, $H_{1/3} \leq 1.25$m), moderate (Douglas 4, $1.25 < H_{1/3} \leq 2.5$m), rough (Douglas 5, $2.5 < H_{1/3} \leq 4$m) and extreme (Douglas 6-9, $H_{1/3} > 4$m). In the North Sea the probability of an extreme sea to occur is very low, being partially sheltered. The last tier was therefore not considered since carrying out a trial in those sea conditions would have been a rare occasion. For calm to rough seas the minimum, most probable and maximum modal periods were extracted. These, combined with the heading and the speed of the trial, returned an estimate of time needed to encounter 100 waves. Since the wave period tends to increase with the wave height, the longest time is usually required in rough seas. It was also found that the most probable wave period of rough seas approximately includes the maximum wave period of a moderate or calm sea. Hence rough sea was adopted as a reference for all the sea states, the advantage being that for a given speed, the same star pattern could be used in different sea states in spite of the actual sea conditions during the trial, whilst being conservative in calm and moderate seas.

A consequence of this approach is that the legs of the star pattern have different lengths depending on the heading. This becomes clear if it is considered that the encounter frequency is a function of the heading and therefore that to encounter the same number of waves less time is required in head sea than it is in following sea. The conceptual star pattern represented in Figure 6 becomes asymmetric (Figure 7). A 9th leg that would be needed to return towards the wave buoy is also added.

The wave encounter period is also a function of the vessel speed. However at speeds as low as 5 knots the effect is almost negligible and a symmetric star with equal time spent at each heading could be adopted. At higher speeds less time is required in head sea and a much longer time is required in quartering and following seas. At 25 knots the difference becomes considerable with head sea runs that require less than 10 minutes against following sea runs that require over 1 hour. Within this time more than 25 nautical miles would be travelled and the distance from the wave buoy would be too large to relate the vessel responses to the wave spectrum measured at the buoy location.

For speeds of 15 knots and above, the star pattern was modified by splitting up quartering and following sea runs into two or more runs. The modified star decreases in diameter and so does the distance from the wave buoy. Its shape for a trial at 25 knots is shown in Figure 8. As a reference, the trial trajectories required 2h50m to complete at 5 knots and 4h30m at 25 knots.

6. DATA ANALISYS

Full-scale trials are invaluable in assessing the response of the actual vessel in a real irregular sea. They also allow investigation of highly dynamic and nonlinear load effects that are challenging to capture with numerical simulations or model tests. For the very same reasons the data analysis is challenging and can significantly affect the outcome.

Two aspects should be considered. The first is the frequency range at which a particular response occurs. Drift is a very low frequency component or even a slowly moving trend; wave-induced responses are related to the wave encounter frequency; and slamming-induced responses tend to occur at higher frequencies. These effects are all superimposed in a raw signal. Frequency-based filters can be applied to the signal to isolate the response of interest.

The second aspect is related to the wave information. A directional wave buoy provides a comprehensive description of the wave environment during the whole trial, but this is relative to the buoy location and not the vessel's. This implies that the data has to be treated with statistical theories. The following sections outline the procedure adopted to assess global loads and local slamming effects.

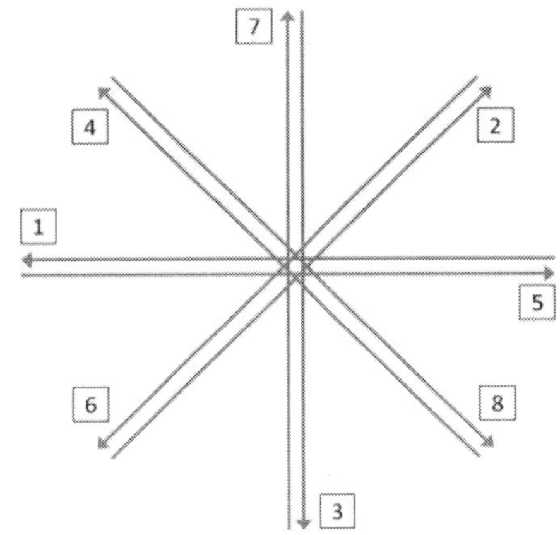

Figure 6: Star pattern with 8 legs at 45° increments.

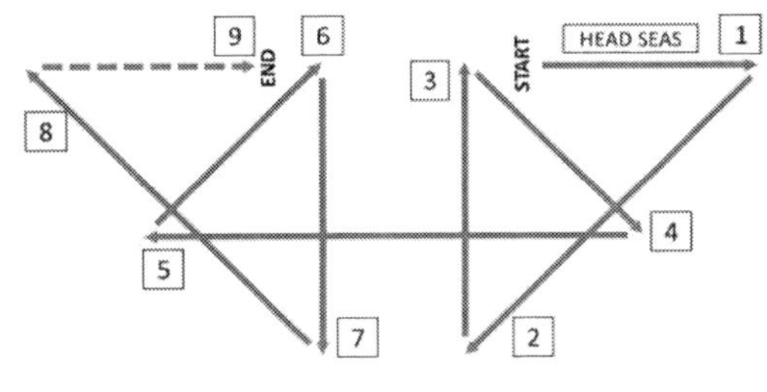

Figure 7: Asymmetric star pattern. The length of each leg is determined as a function of the wave encounter period so that a minimum of 100 wave encounters occurs.

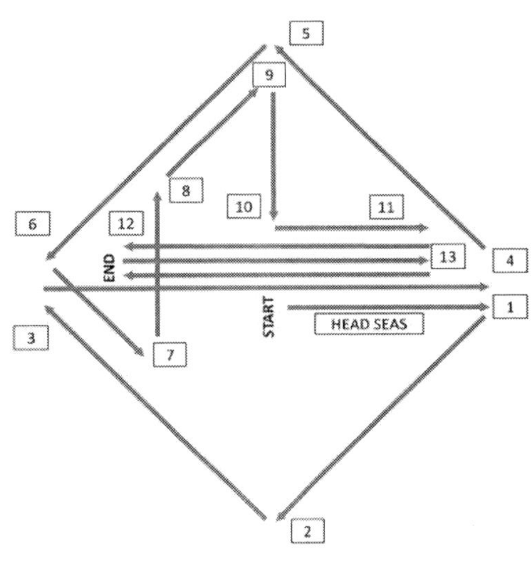

Figure 8: Star pattern adopted at 25 knots. Quartering and following sea runs are split up into two legs to decrease the trial area and the distance from the wave buoy.

6.1 GLOBAL LOADS

Global loads were computed at five longitudinal locations along the hull. At each station three or four strain gauges bonded to primary structural members on the hull bottom and on the deck recorded the hull girder strains (Figure 4). The conversion from strain to hull girder load relies on the classical beam theory and its underlying assumptions. The hull girder is idealised as a beam with material properties and dimensions of the actual vessel. The procedure to recover the vertical bending moment from strain measurements is outlined below. Horizontal bending moment could be estimated in a similar fashion.

The strain gauges were positioned symmetrically with respect to the centreline of the vessel. This is an important expedient to be able to isolate vertical bending strain from a complex loading scenario. For example, due to symmetry, two strain gauges on the deck, port and starboard, will measure same vertical bending strain and opposite horizontal bending strain. Depending on the sign used to combine, or average, the signals of the two strain gauges, either vertical or horizontal bending can be isolated whilst the other neutralised. This approach is at the basis of strain measurement and resembles the theory behind the use of strain gauges in a half or a diagonal Wheatstone Bridge circuit [18]. Whilst a comprehensive explanation is beyond the scope of this paper, it is still noted that superimposed axial loads will be compensated in the successive steps. Superimposed torsional loads will only be partially compensated. However, because of the stiffness and the size of the craft, and because of the orientation of the strain gauges, the magnitude of the parasitic torsional loads is considered negligible.

From the averaging explained above two vertical bending strain signals can be found, one close to the hull bottom and one close to the deck. The arithmetic to recover the magnitude of the vertical bending moment is based on the well-known relationships between strain, stress and load on a bending beam, namely:

$$\sigma = \varepsilon E \tag{1}$$

$$M = \frac{\sigma I}{y} \tag{2}$$

Where ε is the strain, E the modulus of elasticity, σ the stress, I the section modulus of the cross-section, y the distance of the measuring point from the neutral axis and M is the bending moment. Equations 1 and 2 can be found in many engineering textbooks.

The vertical bending strain ε was measured at the deck and at the hull bottom. The modulus of elasticity E was assumed constant and estimated from the material properties of all the reinforcement plies of the cross section. The stress σ could then be calculated from Equation 1 assuming a linear relationship between stress and strain. In order to determine the bending moment M that imparts a stress σ to the structure (Equation 2), values for I and y are also required.

The section modulus I was computed using a finite element global model of the Severn lifeboat built with the marine design software MAESTRO [19]. Details of a first model developed for the study can be found in [20,21]. The model was later refined to improve its accuracy.

The distance y between the measuring point and the neutral axis could have also been retrieved from the finite element model. This was not necessary because the vertical distance between the strain gauges on the hull bottom and those on the deck was known from the technical drawings of the vessel. Equations 1 & 2 can be applied to both the hull bottom and the deck strain signals and they must return the same vertical bending moment. The two equations can then be combined into one that does not depend on y but only on the distance between the strain gauges on the hull bottom and those on the deck. For a positive bending moment in sagging condition Equations 1 & 2 yield to:

$$VBM = \frac{IE}{y_{BD}}(\varepsilon_{Bottom} - \varepsilon_{Deck}) \tag{3}$$

Where ε_{Bottom} is the strain measured at the hull bottom, ε_{Deck} the strain measured at the deck, y_{BD} the vertical distance between the strain gauges at the hull bottom and those at the deck and VBM is the resulting vertical bending moment. From Equation 3 it also becomes clear that axial strain is compensated as it occurs with the same sign at both the bottom and the deck. Torsional strain is only partially compensated since it does occur with the same sign deck and hull bottom but not necessarily with the same magnitude.

A time history of vertical bending moment can be found for each of the five instrumented stations. The time history is made by a wave bending moment and a superimposed dynamic bending moment due to whipping of the hull girder. Noise is also present. Springing is not considered here because of the great stiffness of the hull girder and its high natural frequency.

A band-pass filter is used to isolate the different loading components. The cut off and passband values of the filter were a function of the minimum, mean and maximum wave encounter frequency. An example of how a raw signal can be filtered to isolate wave and whipping components is shown in Figure 9.

6.1 LOCAL SLAMMING PRESSURE

The local deformation of a structural panel due to water pressure was measured on ten panels: six on the hull bottom, two at the bow and two on the fore deck (Figure 4). Except for the deck, the panels instrumented were symmetric port and starboard. Each panel was instrumented with four strain gauges oriented along the longitudinal centreline: two near the supported edges and two at the panel midspan. The strain gauge layout was devised to retrieve both the strain due to highly dynamic loads such as slamming and the strain due to quasi-static wave loads.

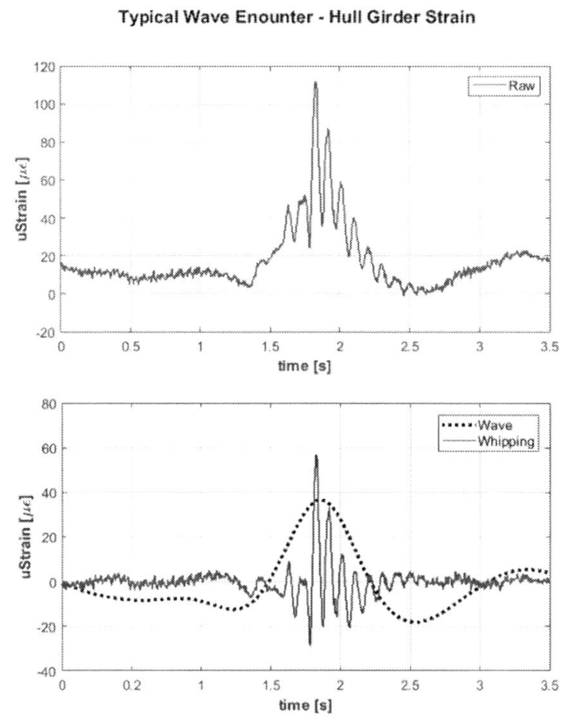

Figure 9: Hull girder strain recorded during a typical wave encounter. Raw signal (top). Filtered signal (bottom) isolating wave and whipping loading.

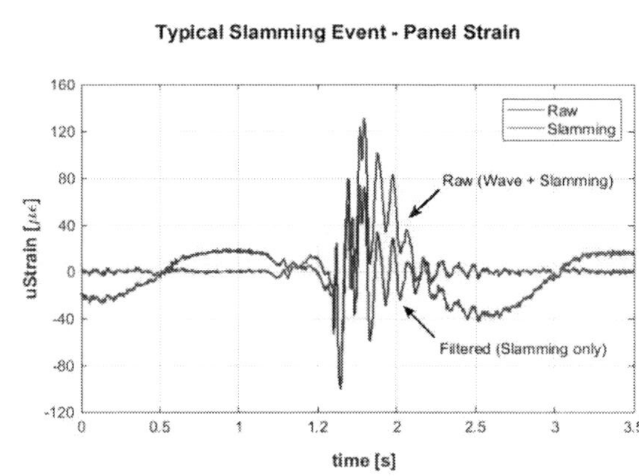

Figure 10: Panel strain recorded during a typical slamming event. Raw signal (wave + slamming) and filtered signal (slamming only).

The computation of the slamming loads relies on the two strain gauges at the panel midspan, positioned close enough that their signals were averaged to obtain one signal. A high-pass filter is applied to isolate the strain component due to slamming, as shown in Figure 10.

A slam is typically characterised by a pressure front travelling rapidly along the panel. Attempting to recover the dynamic pressure field was not practical for the purpose of the study. Such a pressure would then need to be applied dynamically to a finite element model of the vessel. Instead work is being conducted to find an equivalent static uniform pressure. This is the pressure that, if applied to the whole panel, produces at its centre the same strains as those produced by the actual pressure field. This correlation can be found through a local finite element model of the panel under consideration. The analysis is being carried out as linear static and will allow to relate a strain value to a uniform panel pressure value that would have produced it.

6.3 LOCAL WAVE PRESSURE

The computation of the local wave pressure on the hull bottom panels has not been attempted at this stage. However, a brief description of the proposed methodology is given as it explains the strain gauge layout adopted on the structural panels.

The computation of the wave pressure could be carried out in a similar manner as described for slamming pressures by using a local finite element model. A band-pass filter would be applied to isolate the strain due to wave loading from noise and slamming effects. The filtered signal would still embed mainly two components: axial strain due to bending of the hull girder and bending strain of the panel under local wave pressure (torsion of the hull girder is again neglected). These two strain components can hardly be distinguished from one another because they tend to occur at the same frequency: the wave encounter. All four strain gauges become necessary to isolate the local pressure load from the axial load. This can be attempted based on the principle that the four strain gauges will sense approximately the same axial strain due to hull girder bending but different bending strain due to the local pressure on the panel. Beam and plate theories can be applied to estimate the two strain components and isolate one from the other.

7. RESULTS

7.1 TIME DOMAIN STATISTICS

All the trials were first analysed through basic time domain statistics of the measured global wave loads. Maximum, minimum, mean, standard deviation, highest 1/3rd and highest 1/10th values of vertical bending moment were analysed. The purpose of this task was to ensure that the data collected was functional for subsequent analyses. An example of these results is presented in Figure 11, which shows polar diagrams of significant (highest 1/3rd) vertical bending moment peak values for two different trials.

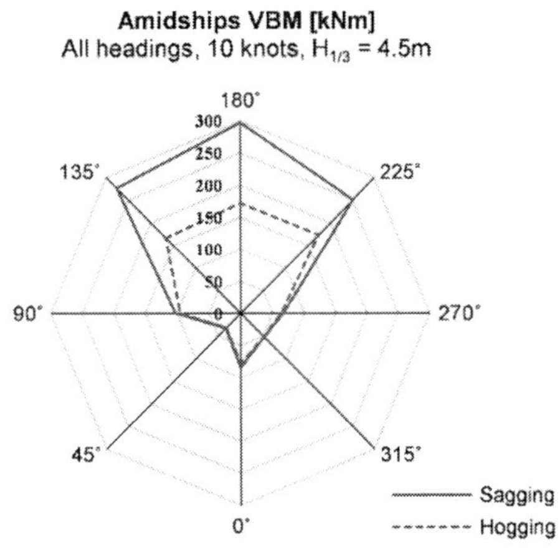

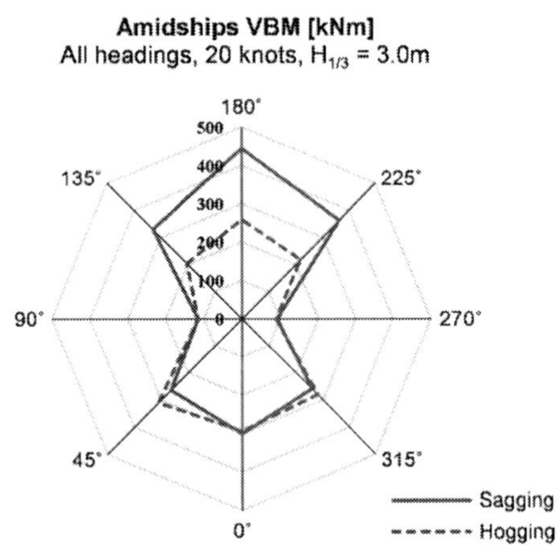

Figure 11: Significant values of Vertical Bending Moment (VBM) peaks for all headings, where 180°is head sea. Trial at 10 knots in 4.5m waves (top). Trial at 20 knots in 3.0m waves (bottom).

The polar plots confirmed what would be reasonably expected. Global wave loads increase with the wave height and the vessel speed. The largest loads are experienced in head seas, followed by bow quartering. Sagging tends to be larger than hogging, with the most significant difference in head seas.

7.2 SPECTRAL ANALYSIS

A spectral analysis was also conducted to calculate the response amplitude operators (RAOs) of vertical bending moment. The response spectrum and the wave encounter spectrum were first calculated. The ordinates of the response spectrum were divided by the ordinates of the wave encounter spectrum at each given frequency. The linear transfer functions were found as the square root of this ratio. The units are those of the ship response amplitude per unit wave amplitude (kNm/m). This analysis was conducted for headings from head to beam seas for which no negative encounter frequencies occur. The RAOs were compared against those obtained numerically from hydrodynamic simulations run with MAESTRO [20,21]. Figure 12 shows the results relative to two different sea trials, conducted at 10 and 20 knots. The magnitude of the response is plotted as a function of the non-dimensional ratio λ/LOA.

Consistent results can be observed for the RAOs computed for different head sea runs during the same trial, which supports the value of the spectral analysis. Where some scatter is present, this could be due to a non-unidirectional sea with swell and wind waves from different compass angles. Partitioning of the sea spectrum could be attempted to isolate the wave systems and treat the ship responses due to each wave system separately.

The longitudinal distribution of vertical bending moment is what would be expected for a conventional hull form. The maximum occurs near amidships, between Frame 5 and Frame 8, and the load decreases towards the aft and forward ends with a trend that resembles a typical longitudinal distribution envelope of vertical bending moment. The RAO trends are also as expected. The peak response occurs for wavelengths of 1 to 2 times the craft length. In shorter and longer waves the response decreases progressively.

At frame 03 the vertical bending moment measured is appreciably smaller than what was predicted numerically. Because of the great stiffness of the hull girder near the stern, only small strains were measured and the wave component did not stand out from the superimposed electrical noise. Part of this discrepancy should be attributed to the impossibility of calculating accurate vertical bending moment values from experimental data.

Frame 10 also shows a discrepancy between numerical and experimental results, visible in Figure 12, but also observed in other RAO plots analysed. At this frame the longitudinal load envelope has its steepest slope dropping from the maximum near amidships to zero at the fore end. What would appear to be a large gap between the two sets of results becomes less of concern when considering the load distribution along the length of the vessel. In fact, it is the longitudinal load envelope that is often more relevant for design.

Another possible source of difference between numerical and experimental data at frames 3 and 10 is the section modulus value adopted for the data analysis. As explained in the previous section these values were estimated from a global finite element model of the vessel, which cannot include all the structural and non-structural members that contribute to the stiffness of the hull girder. The strains measured experimentally are related to the actual section modulus of the cross-section under consideration.

At speeds up to 10-15 knots, the RAOs at Frames 5 and 8 show a satisfactory agreement in trend and magnitude between numerical and experimental results. On one side this suggests the adequacy of the MAESTRO model in predicting the global wave loads on the Severn. On the other side it suggests that the spectral analysis and the computation of the hull girder loads from strain measurement was successful and confirms the quality of the data collected.

At speeds of 15 knots and above the same findings are limited to wavelengths longer than twice the vessel length. Shorter waves seem to produce a much higher response than what was numerically predicted. Because of the dependency to the wavelength, this is likely to be due to two aspects:

- At high speed the transition from displacement to planing regime can affect the vertical bending moment response. This would not be captured by a linear potential flow code like MAESTRO.
- It is possible that wave impacts experienced at high speed cause a nonlinear load response that occurs at frequencies correspondent to approximately $\lambda/LOA \approx 1$. This response is erroneously attributed to waves of that length in the computation of the RAOs, although could have been caused by longer waves. Attempting to filter out this response is not straightforward as it is likely that the linear wave load which occurs at the same frequencies will also be affected.

A global picture of the potential issues associated with numerical values under-predicting the wave loads at high speed will be gained through the analysis of the slamming induced responses. It will then be possible to comment on the contribution of the wave loads, and relative errors, to the total seakeeping load.

8. CONCLUSIONS

An extensive programme of dedicated seakeeping trials with the Severn Class was successfully completed. The data collected allows those dynamic and nonlinear load effects that are challenging to capture with numerical simulations or model tests to be measured under real operational conditions.

The results obtained from the first analyses prove the quality of the data collected and the effectiveness of the tests. They also suggest that the numerical model developed with MAESTRO is capable of predicting the seakeeping loads on the Severn. However, at speeds of 15 knots and above, the numerical simulations seem to underestimate the wave loads sustained. This is likely to be a result of the transition from displacement to planing regime and the possible emergence of nonlinear load effects, neither of which can be captured by a linear code.

Analysis of the highly dynamic seakeeping loads at a global and local level – whipping of the hull girder and local slamming pressures – is ongoing. The results will make it possible to account for these load effects in the calculation of the structural response of the Severn. It will also be possible to comment on the relative magnitude of wave and slamming loads and on the error associated with their estimation.

9. ACKNOWLEDGEMENTS

The sea trials were carried out as part of a project funded by the Royal National Lifeboat Institution.

This paper also forms part of a PhD Thesis, which will be submitted in 2018, supervised by Prof Robert S. Dow, Prof Richard W. Birmingham and Dr Simon Benson.

The authors would like to thank the RNLI crews of Tynemouth Lifeboat Station who made the sea trials possible.

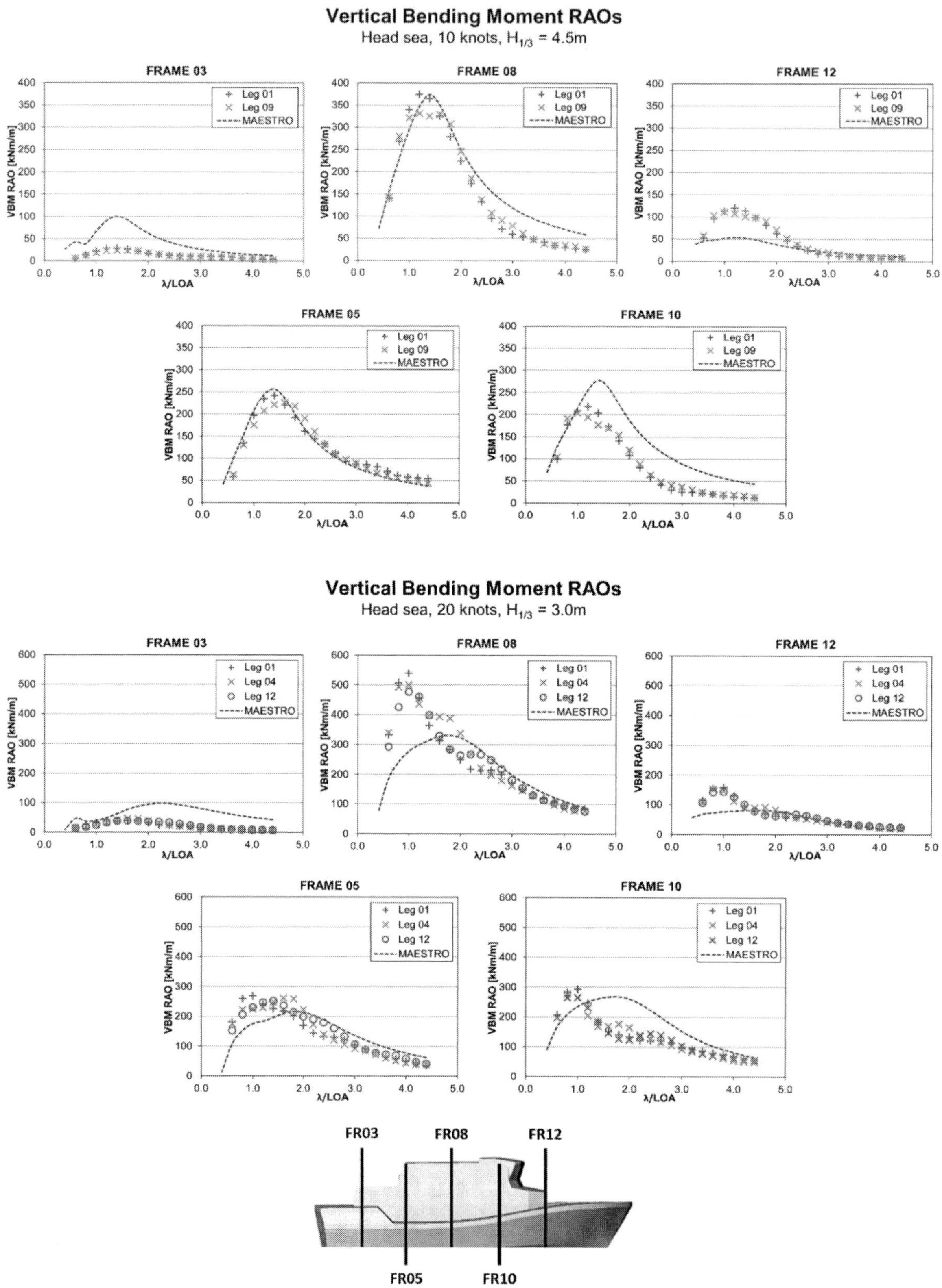

Figure 12: Vertical Bending Moment RAOs in head sea. Experimental data from two different sea trials: at 10 knots in 4.5m waves (top) and at 20 knots in 3.0m waves (bottom). Numerical data from simulations run with MAESTRO [20,21]. The magnitude of the response is plotted against the ratio wavelength/ship length (λ/LOA).

10. REFERENCES

[1] Vishay Precision Group, "Strain Gage Instrumentation," 2016. [Online]. Available: http://www.vishaypg.com/docs/50002/50002.pdf. [Accessed: 20-Feb-2016].

[2] O. M. Faltinsen, *Hydrodynamics of High-Speed Marine Vehicles*. Cambridge University Press, 2006.

[3] Vishay Precision Group, "System 7000 Datasheet." 02-Jan-2012.

[4] A. R. J. M. Lloyd, *Seakeeping: Ship Behaviour in Rough Weather*. Ellis Horwood, 1989.

[5] E. A. Tannuri, J. V. Sparano, A. N. Simos, and J. J. Da Cruz, "Estimating Directional Wave Spectrum Based on Stationary Ship Motion Measurements," *Appl. Ocean Res.*, vol. 25, no. 5, pp. 243–261, Oct. 2003.

[6] C. Drouet, N. Cellier, J. Raymond, and D. Martigny, "Sea State Estimation Based on Ship Motions Measurements and Data Fusion," presented at the 32nd International Conference on Ocean, Offshore and Arctic Engineering OMAE2013, Nantes, France, 2013.

[7] U. D. Nielsen, "Estimations of On-Site Directional Wave Spectra from Measured Ship Responses," *Mar. Struct.*, vol. 19, no. 1, pp. 33–69, Jan. 2006.

[8] M. C. Johnson, "Improvements in the Conduct and Interpretation of Ship Seakeeping Trials," PhD Thesis, University of Southampton, Southampton, UK, 2004.

[9] Datawell B.V., "Datawell Waverider Reference Manual for DWR-MkIII and DWR-G." 01-Dec-2014.

[10] J. J. de Vries, J. Waldron, and V. Cunningham, "Field Tests of the New Datawell DWR-G GPS Wave Buoy," *Sea Technol.*, no. 44, pp. 50–55, Dec. 2003.

[11] G. Jeans, I. Bellamy, J. Jan de Vries, and P. van Weert, "Sea Trial of the New Datawell GPS Directional Waverider," presented at the 7th Working Conference on Current Measurement Technology, 2003, pp. 145–147.

[12] National Oceanic and Atmospheric Administration, "Sea State Photographs for Determining Wind Speed." National Oceanic and Atmospheric Administration, U.S. Department of Commerce, May-1987.

[13] Met Office, "Fact Sheet 6 - The Beaufort Scale," 2010.

[14] E. V. Lewis, *Principles of Naval Architecture*, vol. 1 Stability and Strength. The Society of Naval Architects and Marine Engineers, 1988.

[15] S. L. Bales, W. T. Lee, and J. M. Voelker, "Standardized Wave and Wind Environments for NATO Operational Areas," David W. Taylor Naval Ship R&D Center, DTNSRDC/SPD-0919-01, Jul. 1981.

[16] W. T. Lee, "Global Wave Statistics for Structural Design Assessments," Naval Surface Warfare Center, NSWCCD-HD-1048-01, Oct. 1995.

[17] W. T. Lee and S. L. Bales, "Environmental Data for Design of Marine Vehicles," presented at the Ship Structure Symposium '84, 1984.

[18] H. Karl, *An Introduction to Measurements using Strain Gages*. Hottinger Baldwin Messtechnik GmbH, 1989.

[19] MAESTRO Version 11.5.0, "MAESTRO User Manual." Advanced Marine Technology Center, DRS Technologies Inc., 2017.

[20] F. Prini *et al.*, "Motions and Loads of a High-Speed Craft in Regular Waves: Prediction and Analysis," presented at the 24th International HISWA Symposium on Yacht Design and Yacht Construction, Amsterdam, NL, 2016, pp. 1–14.

[21] F. Prini *et al.*, "Seakeeping Analysis of a High-Speed Search and Rescue Craft by Linear Potential Theory," presented at the International Conference on Lightweight Design of Marine Structures, Glasgow, UK, 2015, pp. 87–96.

11. AUTHORS BIOGRAPHY

Federico Prini is Research Associate in the Marine, Offshore and Subsea Technology Group within the School of Engineering at Newcastle University. He is working on the research project reported in this paper with the Royal National Lifeboat Institution (RNLI) and Lloyd's Register. He gained internship experience with the shipbuilding company Ferretti Group. With a bachelor's degree in Pleasure Craft Engineering and a Master in Small Craft Design he is currently registered for a PhD in marine hydrodynamics and structures.

Simon Benson holds the current position of Lecturer in Naval Architecture at Newcastle University and academic director of the Marine Propulsion Laboratory. He is responsible for teaching and research in marine hydrodynamics and structures. His previous experience is in industry working for BMT Fluid Mechanics and IHC Engineering Business before completing a PhD investigating the structural performance of aluminium high speed ships. His PhD and post-doctoral work were sponsored by the Office of Naval Research. The work has resulted in the release of a progressive collapse code (ProColl) which is now incorporated into the commercial ship structural analysis program MAESTRO and is used by the US Navy amongst others. He was appointed as a lecturer at Newcastle University in October 2012. He became director of the marine propulsion lab in October 2017.

Richard Birmingham is Professor of Small Craft Design at Newcastle University. His areas of research include safety and sustainability in the context of small craft design, and he has worked collaboratively with the RNLI for 20 years. He is the Principal Investigator for the work reported in this paper.

Robert S Dow is Professor in Marine Structures and Head of the Marine, Offshore and Subsea Technology Group within the School of Engineering at Newcastle University. He is responsible for teaching and research in marine hydrodynamics and structures. He has extensive experience of full-scale sea trials with measurement of structural loads and response. He is also Co-Investigator for the work reported in this paper.

Luke J Ferguson holds the current position of Group Leader - Shannon Fit-out at the RNLI. He is responsible for the build and fit-out of the Shannon Class All-Weather Lifeboat at The All-Weather Lifeboat Centre in Poole, Dorset. His previous experience includes electrical/electronic engineering and more recently mechanical design analysis and part optimisation whilst studying for a master's degree at Bournemouth University.

Peter J Sheppard holds the current position of Senior Naval Architect at the RNLI. He is responsible for Whole Body Vibration and seating trials as well as trouble shooting composite processing, materials, repairs and testing. His previous experience includes managing all RNLI new-build work at Green Marine including the construction of Severn, Trent and Tamar classes.

Holly Phillips holds the current position of Principal Naval Architect at the Royal National Lifeboat Institution. She forms part of the RNLI's Design Authority working group and manages the naval architecture team responsible for the structural and stability aspects of the RNLI's fleet of all weather and inshore lifeboats, lifeguard equipment and other assets. Her previous experience includes that of Project Manager responsible for the design and development of the E-Class lifeboats for the River Thames.

Michael C Johnson is a senior specialist in hydrodynamics at Lloyd's Register. He is responsible for LR's internal developments in the seakeeping and loads field and is an experienced user of these codes for LR's research programme and for consultancy work. He has long standing personal interest on the wave environment side of the problem, both in the practical measurement of waves and the representation in computation assessments, such as analysis techniques and the contribution of wave spreading and directionality. He chairs the CRS 'CRISM' working group making a long term auto-logging campaign on two active containerships. He has conducted many model test experiments and full scale seakeeping trials, including as seakeeping trials officer for the trimaran research vessel RV Triton.

Jesus Mediavilla Varas holds the current position of Computational Mechanics and Valorisation Manager at the Materials Innovation Institute M2i. His involvement with the work reported in this paper started whilst he was holding the position of Project Manager and Lead Specialist in Strategic Research at Lloyd's Register.

Spyros Hirdaris holds the current position of Visiting Scholar at Aalto University (Finland). His current role involves teaching and research in ship dynamics. From 2004 to January 2018 he has worked for Lloyd's Register Technology & Innovation at UK and the Far East.

RESCUE BOATS ON CATAMARAN PLATFORM: CASE STUDIES

A Nazarov, Albatross Marine Design, Thailand

SUMMARY

Catamaran platform provides certain benefits for some rescue tasks, but unfortunately is not widely applied for such applications. Numbers of rescue catamarans with paramedical, dive support and firefighting capabilities have been developed by Albatross Marine Design team during recent years. Based on such experience, the paper presents catamaran design specifics related to general dimensions, architecture, performance and structural design of these craft. Parametric approach to dimensioning a typical high-speed catamaran craft with payload, performance and seakeeping restrictions is proposed. Perspective areas of catamaran applications are discussed; new concepts and their potential applications are featured.

Figure 1: AS14 rescue/paramedical catamaran designed by AMD for operation at Volga river estuary. The boat was featured in Significant Small Ships 2014 by RINA.

1. INTRODUCTION

Catamaran platform is increasingly applied in boat design practice for recreational, passenger, wind farm support and other craft. Some niche exists for application of catamaran scheme for rescue boats as well.

During recent years, Albatross Marine Design (AMD) has developed number of small catamaran craft projects, including boats intended for rescue and related applications. These craft can cover some of the following functions:

- Search and rescue
- Paramedical treatment and medical transport
- Patrol and law enforcement
- Firefighting
- Dive support
- Eco and spill monitoring
- Command centre
- Personnel and equipment transport

Based on design, construction and operation experiences, the author makes recommendations on design specifics and perspectives of catamarans.

2. CATAMARAN PROS AND CONS

With wide beam catamarans offer more usable space for restricted length. These boats are stable and seaworthy platforms, with plenty of space to fit range of rescue and paramedical equipment. Catamarans also offer improved safety: high initial and damaged stability, ample freeboard, duplicate propulsion units and systems, etc.

Catamaran platforms allows staying inside 'magic length' of 24m within 'small craft' classification, resulting savings on equipment and class maintenance.

However, 'more space' approach can turn into serious disadvantage: given same length of craft, average catamaran is heavier than monohull, due to greater area of structures and usually exaggerated customer's payload requirements on wider deck. These extra structures and equipment lead to higher construction cost of catamarans.

One of important limiting factors is higher expertise level required to design catamaran craft, which often means higher risk for the whole project. This is possibly the reason, why despite functional advantages, catamaran craft are still not fully welcomed by most of security and rescue agencies, especially in government tenders. Thus, conceptual designs of catamarans are often rejected due to lack of trust in this type of craft.

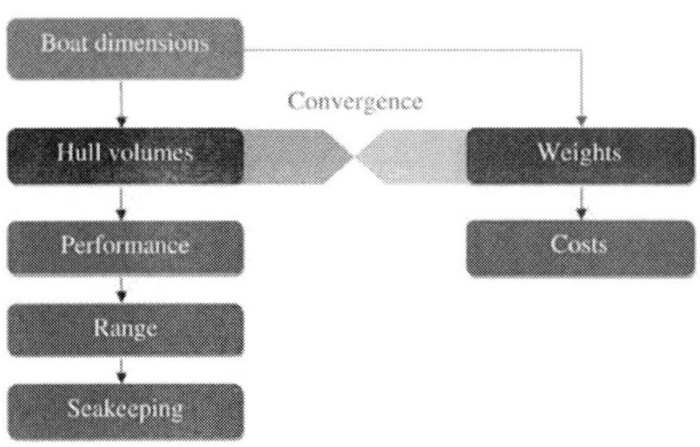

Figure 2: Block diagram of parametric design method

It should be said that self-righting capability, which is often seen as 'a must' for rescue craft, is hardly possible to achieve on a catamaran. But research [7] shows that for coastal and inshore applications a) the probability of capsize is negligent and b) capsizing conditions are related to beam of the craft. Thus 'prevent capsize' rather than 'provide capsize recovery' approach can be justified for catamarans, due to their 'beamy nature'.

Taking all together, catamaran platforms are justified and proved to be efficient for high-speed moderate sea state applications, for space restricted applications with high stability demand. This envelope would include many rescue craft needs.

To make catamaran design efficient, careful analysis should be performed at all design stages with application of existing engineering methods and experience of previous designs.

3. GENERAL DESIGN CONSIDERATIONS

3.1 PARAMETRIC DESIGN APPROACH

General approach used by AMD is parametric design analysis at early stages. The approach allows filtering out unfavourable solutions, such as heavier catamarans of smaller size, for given payload.

Parametric design method is published by author in [4] and was later modified for catamaran shapes. Usually, parametric study is undertaken at pre-drawing stage, and intermediate where results are presented to the client in terms of efficiency and feasibility. Parametric approach is based on reaching convergence of hull volumes and weights (fig.2); base system of design equations describing a high-speed craft can be written as following:

$$\left.\begin{aligned} &\sum v_i \le v_A \\ &M = \rho g V \\ &v_S \ge [v_S]_R \\ &r \ge [r]_R \\ &\frac{a_{CG}}{g} \le \left[\frac{a_{CG}}{g}\right]_{\max} \\ &S \to \min \end{aligned}\right\}$$

The first equation is relation of sum of required volumes v_i and available volume v_A; second equation is balance of craft mass M and volume displacement V; third and fourth are comparisons of design speed v_S and range r with required speed $[v_S]_R$ and range $[r]_R$. Fifth equation specifies condition of seakeeping as not exceeding certain acceleration level a_{CG}/g. Last equation is criteria of minimization of construction cost S, though operational costs might be considered for complete analysis. In practice, parametric analysis is often 'anchored' to existing designs for higher degree of accuracy of predictions.

Example of design parameters generated by the method is given Table 1. In practice, the design can be restricted by dimensions, draught, transportability requirements, demihull beam for installation and service of engines, etc.

3.2 STABILITY

At pre-design stage, stability is excluded from preliminary study, as for catamarans intact stability is excessive and will normally comply with high margin. Damaged stability of catamarans below 24m is governed by engine room flooding as worst possible case (fig.3).

In AMD's practice, stability compliance of catamarans is checked versus IMO HSC Code [1] or ISO12217-1 [2]

Table 1

PLANING CATAMARAN - PARAMETRIC DESIGN Rescue Cat

GEOMETRY:

Length of hull, m	L_H	15	
Length of WL, m	L_{WL}	14,25	
Beam of hull, m	B_H	5	
Beam of chine, m	B_C	1,54	
Canoe body draft, m	T_C	0,7	
Hull depth, m	D	2	
Volume displacement, m^3	V	20,92	From TW:
Weight displacement, kg	DSPL	21441	21442
Immersion, kg/cm	q	176	

SHAPE FACTORS:

Transom deadrise angle, deg	β	12	
Block coefficient	C_B	0,68	For one hull
Midship area coefficient	C_M	0,88	For one hull
Waterplane area coefficient	C_{WP}	0,80	For one hull
Static load factor	C_Δ	0,72	For one hull

PERFORMANCE:

Load condition		LDC	MOC	Cruising
Boat speed, kts	v_S	45,0	51,9	30,5
Boat speed (LDC), m/s		23,13	26,68	15,69
Froude number, Fn_L		1,96	2,26	1,33
Froude number, Fn_V		4,45	5,13	3,02
RPM, %		100	100	70
Propulsive efficiency	η	0,55		
K	K	210		
Effective power (EHP), kW	P_E	857		
Power demand (SHP), kW	P_S	1559		
Power demand (SHP), h.p.	P_S	2118		974
Power margin, %		5		
Derating factor, %		0		
Required power, h.p.		2223		1023

FUEL CONSUMPTION / RANGE:

Load condition		LDC	MOC	Cruising
Specific fuel consumption, g/h.p./h	q_F	160		
Fuel density, kg/m^3	ρ_F	850		
Fuel consumption per nautical mile, L/nM		9,30	8,06	6,31
Range, nM	R	358	413	528

SEAKEEPING:

Deadrise angle at STN5, deg	β_S	17,1		
Vertical accelerations, 1/g	a_{CG}/g			
Sea states:		LDC	MOC	Cruising
SS2		0,58		0,27
SS3		1,18		0,54
SS4		2,22		1,02
SS5		3,63		1,67

WEIGHTS:

Structure:		
Structural mass factor, kg/m^2	m_H	35
Superstructure factor	k_S	1,1
Mass of structure, kg	M_H	5775
Interior & outfit:		
Interior mass factor, kg/m^2	m_I	5
Mass of interior & outfit, kg	M_I	750
Machinery:		
Number of engines	n	2
Engine mass factor	m_E	1,3
Mass of engines	M_E	2890
Drive type factor	k_D	1,3
Machinery mass factor	k_M	1,5
Mass of machinery	M_M	4769
Deck equipment:		
Deck equipment mass factor	m_D	0,35
Mass of deck equipment, kg	M_D	524
Systems:		
Systems mass factor	k_{SYS}	0,35
Mass of systems, kg	M_{SYS}	524
Electrical:		
Electrical system mass factor	k_{EL}	0,35
Mass of electrical system, kg	M_{EL}	524
Other:		
Additional equipment	M_A	2500
Weight margin	M_M	0
Light craft mass, kg	M_{LCC}	15367
Safety equipment	M_S	156
Non-consumable stores	M_{NCS}	0
Crew at MOC	n_{MOC}	2
Mass of crew at MOC	M_P	200
Minimum operation mass, kg	M_{MOC}	15723
Crew/passengers/personnel:		
Number of crew/passengers/personell, kg	n	6
Mass of one crew/pasenger	m_P	100
Mass of crew/pasengers, kg	M_P	600
Fuel:		
Fuel capacity factor, L/h.p.		2
Fuel tanks capacity (estimated), L		4447
Fuel tanks capacity (used), L		3500
Mass of fuel, kg	M_F	2975
Other liquids:		
Fresh water capacity (used), L	M_F	300
Waste tank capacity (used), L	M_W	200
Other liquids	M_L	0
Cargo	M_{CRG}	2000
Payload	M_{MTL}	6075
Loaded displacement mass, kg	M_{LDC}	21442

Structural mass factors, kg/m^3

FRP sandwich	25...40
FRP single skin	40...55
Aluminum	35...60

Interior & outfit mass factors, kg/m^3:

Luxury yachts	18...25
Sport cruisers, cabin cruisers	7...20
Patrol boats, special craft	5...12
Open/console boats	5...10

Engine mass factors, kg/h.p.:

Diesel engines	1.6...1.9
Gasoline engines	1.3...1.5

Machinery mass factors:

All types	1.35...1.7

Drive type factors:

Shaft drives	1
Sterndrives	0.9
Jet drives	1.3

Deck equipment, systems, electrical mass factors:

	Deck	Systems	Electrical
Min	0,3	0,3	0,3
Average	0,37	0,39	0,35
Max	0,5	0,5	0,5

Crew at MOC:

L_H<8m	1
L_H=8...16m	2
L_H>16m	3

Mass of crew/passengers, kg

Passenger/crew	75
Passenger with luggage; diver	100
Personel with ammunition	120

Fuel capacity factor, L/h.p.:

Diesel engines, pleasure craft	1.7...2.5
Gasoline engines	1.3...2.5

COSTS:

Cost factos:		
Structure, THB/kg	s_S	1 000
Interior/outfit, THB/kg	s_I	640
Machinery/systems, THB/kg	s_{MS}	1 400
Construction cost estimates:		
Structure, THB	S_S	5 775 000
Interior/outfit, THB	S_I	480 000
Machinery/systems, THB	S_{MS}	8 878 750
Construction cost:	S	15 133 750

requirements, where the first is preferred for number of reasons [5].

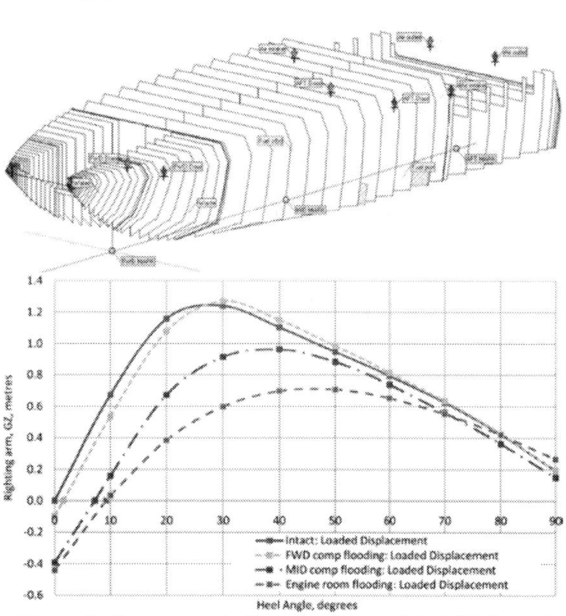

Figure 3: Catamaran hull geometry model (AS14) used for stability calculations (above), and resulting stability diagrams for intact and damaged conditions (below).

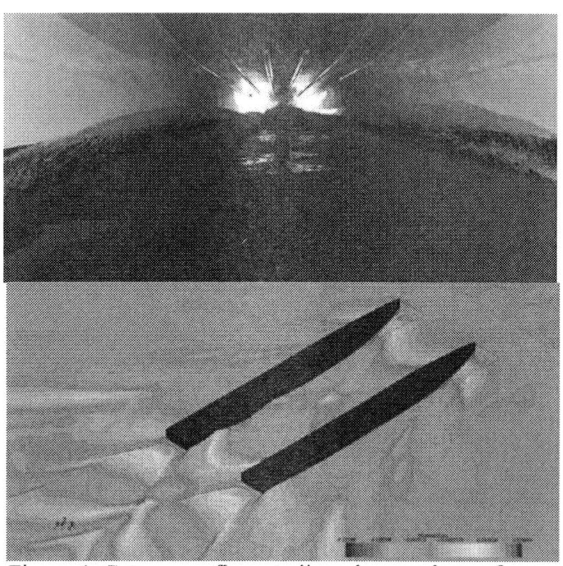

Figure 4: Catamaran flow studies: above - photo of wave making between hulls of 'symmetrical' catamaran; below - CFD for 24m symmetrical hull at 10kts.

3.3 PERFORMANCE

'High performance' is often referred as one of attractive features of catamarans. Properly designed catamaran can theoretically possess reduced hydrodynamic resistance compared to similar monohull craft, in certain range of speeds and displacements. In most of real-life cases, catamaran will have higher resistance - given same length of craft, average catamaran is heavier than mono, considering greater area of structures and usually exaggerated customer's payload requirements on wider deck. From hydrodynamic considerations, catamaran platform makes sense if slenderness ratio for demihull $L/(V/2)^{1/3} \geq 6$, where V is craft's volume displacement, especially at semi-planing mode. For some applications where stability concerns are priority, heavier catamarans can be accepted.

In practice, real advantage of catamarans is their ability to maintain higher speed at given sea state, due to reduced vertical accelerations [6].

Performance of small catamarans is in detail reviewed by the author in number of papers [5], where recommendations on hull shape selection and resistance prediction methods are provided. However, design development is often based on results of sea trials and operation of previously built craft, these results allow verification of performance predictions with corrections for interaction with propulsion system. In recent years, CFD methods have been increasingly applied by AMD for optimisation of hull shapes and design of details such as propeller pockets.

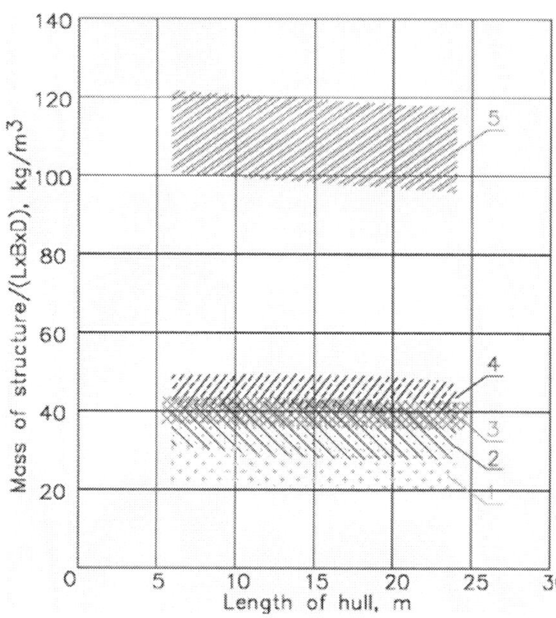

Figure 5: Structural mass factors for catamarans 1 – composite, displacement catamarans with small superstructure, sailing catamarans; 2 – composite, high-speed catamarans with small superstructure; 3 – aluminium catamarans with small superstructure; 4 – composite, catamarans with large superstructure and/or flybridge; 5 – steel catamarans.

3.4 STRUCTURAL DESIGN

Structural design of catamarans should be concentrated around creating of weight-efficient structure. Being 'overbuilt', catamarans with their excessive area of structures are becoming heavy and lose any performance advantages. AMD has performed experimental study of vertical accelerations and associated impact loads on number of catamarans [6]. Some useful conclusions are:

- Structural design loads for high-speed craft are based on 1960s' research with relatively inferior impact recording technique.
- For craft operated in semi-planing and early-planing modes, existing methods used by classification societies correctly predict pressures on the forebody, but seriously overestimate bottom pressures at the aft part of the hull.
- Data-logging of accelerations and bottom pressures is a valuable source of information to establish operational limitations for existing craft, and to enhance design procedures for future designs.

Figure 6: Construction methods for catamaran craft: AS14 aluminium hull (above); infusion process for ASV1500 composite hull (below).

In catamaran design practice, scantling rules producing heavy structures should be avoided. For instance, rules prescribing minimum thickness of skins for composite sandwich structures are not appropriate for most of smaller size catamarans [3].

For catamarans below 24m, composite materials result significantly lighter structures compared to aluminium and especially steel. On fig.5 the cubic mass factor is presented, based on hull length *L*, overalll beam *B* hull depth *D*. This simplified factor is intended to be used in parametric model, at pre-drawing stage where configuration and dimensions of superstructures are not yet defined. The statistics on fig.5 represents AMD's projects with 'weight-savvy' approach in mind (fig.6).

In AMD's practice, carbon fibre is increasingly used for composite catamarans, such as SM18 (fig.9), where weight saving comprises over 20%. The reduction of structural mass is utilized into corresponding increase of payload capabilities.

Figure 7: R38 rescue/paramedical catamaran
Particulars: length overall 12.26m, beam 3.81, draught 0.93, displacement light/full load 5.8/9.0t; power 2x315HP, speed 38kts, hull material - FRP

4. R38 DESIGN

The first rescue catamaran designed by AMD was R38 (fig.6) - 12m composite craft developed for tsunami relief fund – an organization assisting recovery of Asian countries after 2004 disaster. The boat was built in Thailand in 2007 in fiberglass, and donated to Thai government. The main purpose of the craft was providing paramedical assistance on remote islands, where medical treatment was inexistent. With top speed of 38kts at sea state 3, the craft is capable to promptly reach remote locations on Andaman coast. In case of on-water rescue, the craft is capable to take up 30-50 persons on board, and proceed slowly to safe location. During its first year of service, the boat provided assistance to about 50 persons.

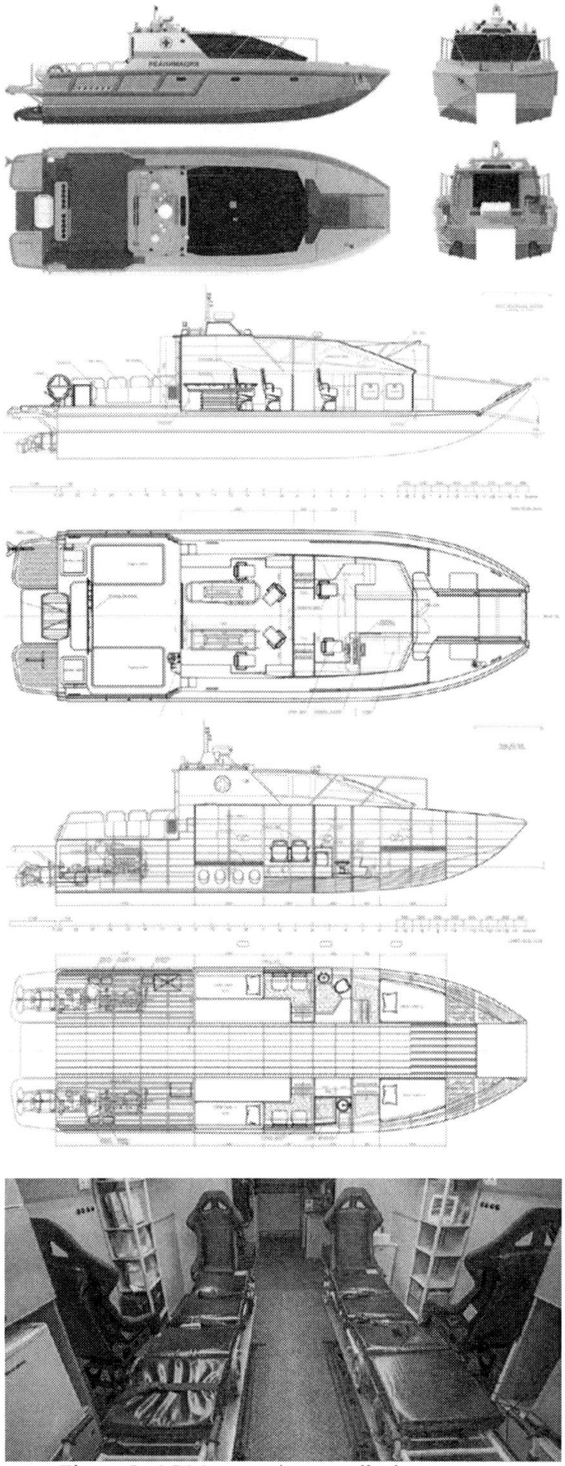

Figure 8: AS14 rescue/paramedical catamaran. Particulars: length overall 15.34m, beam 4.89, draught 0.80, displacement light/full load 14.5/19.0t; power 2x850HP, speed 45kts, hull material - aluminium.

5. AS14 DESIGN

The further development of R38 was AS14 design (fig.1 and 8) – larger catamaran designed in aluminium for client in Russia. The craft is intended for rescue and paramedical services in delta of Volga river. Features of this craft include:

- Larger medical compartment, two stretchers with shock absorbing.
- Bow ramp and opening front window of pilothouse enable beach loading into medical compartment.
- Provision for crane, dive tanks, rescue gate, etc.
- The craft is capable to operate in thin ice, at slow speed.
- Accommodation spaces in demihulls, including cabins, shower, pantry.

The AS14 is delivering 45kts fully loaded and is fully serviceable to sea state 4. Hull shape of AS14 is 'split hull' type, which enables easy fit of bow ramp.

The craft was designed in 2013 and launched in 2014. The catamaran scheme proved to be advantageous and operator's comments on the craft are very positive. In next few years, AMD has received inquiries for similar designs, both in aluminium and composite.

6. PERSPECTIVE PROJECTS

6.1 SM180 'BELUGA' PROJECT

The client's requirement for this project was catamaran for coastal rescue and fire fighting in Middle East countries. Initially, the requirement was 15m catamaran, however with increasing equipment and payload the craft finally became 18m with 40+ tons of full load displacement.

Figure 9: SM18 rescue catamaran with fire fighting capability. Design particulars: length overall 17.91m, beam 6.51, draught 1.00, displacement light/full load 32/40t; power 2x1600HP, speed 40kts, hull material - composite, classification LR SSC.

Architecture of the craft is walk-around design with high superstructure and flybridge. The aft deck is open working area and can be used to carry RIBs or additional equipment, also for dive support operations. Accommodations are provided for crew and personnel in the demihulls, with main deck space used for steering station and saloon/office/paramedical zones.

The biggest challenge for the project was reduction of mass of structure which is achieved by using of carbon in laminate, in absolutely all composite parts of the craft. The craft is currently under construction with launching expected in 2018.

6.2 M085EC1 CATAMARAN

The main purpose of the craft is rescue operations, first aid assistance and fire fighting on inland waterways of Thailand, where such services are inexistent. The concept is developed in aluminium alloy. The craft will feature rescue equipment, one stationary fire pump and also portable fire pumps. Bow ramp is provided for easy unloading of equipment.

Figure 10: M085EC1 rescue catamaran with fire fighting capability. Design particulars: length overall 10.24m, beam 4.22, draught 0.56, displacement light/full load 6.0/8.2t; power 2x160HP, speed 17kts, hull material - aluminium.

7. CONCLUSION

Catamaran platform offers obvious functional benefits for rescue applications, for craft designed for coastal operation, paramedical treatment, firefighting and other similar functions.

However, the advantages are not always appreciated by some operators and government agencies due to 'monohull' stereotypes and formal tendering procedure limitations. Awareness of all parties is required to change the situation towards more optimum craft types.

Based on experience of realised projects, we at AMD are confident of advantages catamaran craft and are developing new concepts of such boats for rescue applications.

8. REFERENCES

1. International Code of Safety for High Speed Craft (2000 HSC Code) – IMO, 2008 Edition
2. ISO-12217 - Small Craft Stability and Buoyancy Assessment and Categorization.
3. Nazarov A. Composite High-Speed Craft: Practical Design Approaches// The 4th Chesapeake Power Boat Symposium, Annapolis, Maryland, 2014.
4. Nazarov A. On Application of Parametric Method for Design of Planing Craft //3rd Chesapeake Power Boat Symposium, USA – June 2012.
5. Nazarov A. Small Catamarans: Design Approaches and Case Studies//International Journal of Small Craft Technology, RINA, Vol.157, Part B1 2015.
6. Nazarov A. Suebyiw P., Piamalung A. Experimental assessment of Impact Loads on Catamaran Structures// Design & Operation of Wind Farm Support Vessels, 28-29 January 2015, London, UK – p.75-84.
7. Nazarov A., Suebyiw P., Leeprasert A., Piamalung A., Surasorn M. Small Patrol Boats: Design for Self-Righting//International Jounrnal of Small Craft Technology, RINA, Vol.155 Part B2 2013. p.71-80.

9. AUTHORS BIOGRAPHY

Dr.Albert Nazarov holds the degree in Naval Architecture from Sevastopol National Technical University (1996) and Ph.D. in Ship Dynamics from Odessa National Maritime Academy (2004). His current position is managing director at Albatross Marine Design with experience in construction, design and research of small and high-speed craft.

Numerical Investigation of Double Continuous Welding and alternatives in Bottom Shell Plates of Aluminium High-Speed, Mono-hull Craft.

Vimal Kumar, Plan Approval Specialist, Lloyd's Register Singapore
Dr Burak Can Cerik, Assistant Professor, School of Marine Science and Technology, Newcastle University

SUMMARY

In this study; numerical strength investigation, of three welding configurations (DC, CI and SI) used in bottom shell of aluminium high-speed mono-hull craft, is carried out. Applied load is mainly bottom impact pressure, which exceeds combined load of hydrostatic and hydrodynamic loads and mainly dependent on vertical acceleration at L_{CG}, are estimated based on the LR SSC Rules. In-plane compressive loads due to global bending of the craft are considered insignificant and ignored in the study. Structures including the weld metals are modelled as solid geometries. Stresses in whole structures including weld-metals and weld-metals exclusively are estimated with the help of Ansys Mechanical 17.2. Further, simplified fatigue approach is used to estimate fatigue life of structures in different weld configurations. The study concludes, where predominant loads are bottom impact pressure, stress variation in different weld configurations is minimal. It also observes; based on appropriate S-N curves for different welding configurations, fatigue life of structures with intermittent welding is significantly lower than that of with DC welding configuration.

Keywords: High-speed craft, Aluminium, Intermittent Weld, Stress, Fatigue life

NOMENCLATURE

Symbol	Definition
d	Weld spacing (mm)
D	Cumulative fatigue damage
F_∇	Volumetric speed number
K	Stress concentration factors
L_{CG}	Longitudinal centre of gravity
L_{OA}	Length over all (m)
L_R	Rule length (m)
L_{wl}	Summer waterline length (m)
N_R	Reference life (Cycles)
N_T	Design life (Cycles)
P_b	Plate width (mm)
P_h	Hydrostatic pressure (KPa)
P_w	Hydrodynamic wave pressure distribution due to relative motion (KPa)
s	Weld length (mm
S	Random variable denoting stress range
S_Q	Stress range at which the slope of the S-N curve changes
S_R	Reference stress range (MPa)
T	Plate thickness (mm)
T_r	Throat thickness (mm)
T_x	Local draught (m)
U	Displacement along X-Axis
V	Displacement along Y-Axis
V_s	Allowable speed (Knots)
W	Displacement along Z-Axis
X_a	Distance from aft end at static load waterline (m)
Y	Weibull shape parameter
z_k	Vertical distance of the underside of the keel (m)
α	Angle of attack (degree)
Γ	Taylor Quotient
δ	Weibull scale parameter
∇	Moulded displacement (m^3)
μ	Toe angle (Radians)
$\Delta\sigma$hot-spot	Hot-spot stress range (MPa)
$\Delta\sigma$nominal	Nominal stress range (MPa)
θ_x	Rotation about X-Axis
θ_y	Rotation about Y-Axis
θ_z	Rotation about Z-Axis
λ	Coefficient for end supports
ρ	Toe radius (mm)
υ	Angular mismatch (Radians)
ω	Wave frequency (Hz)
ω_e	Wave encounter frequency (Hz)

LIST OF ABBREVIATIONS

ABS	American Bureau of Shipping
CI	Chain intermittent fillet welding
DC	Double continuous fillet welding
DNV	Det Norske Veritas□
HAZ	Heat affected zone
LR	Lloyd's Register
SI	Staggered intermittent fillet welding
SSC	Special service craft

1. INTRODUCTION

Welding is the most common joining technique in the shipbuilding industry despite being reasons for various defects and even structural failures. There had been many accidents, associated with problems caused mainly due to welding. These have resulted, not only in the loss of properties and environment but also have been fatal. Most infamous accident, the collapse of Norwegian semi-submersible drilling rig *Alexander L. Kielland* attributes to the fatigue failure in weld areas and considered as the worst disaster in Norwegian offshore history. Currently, problems associated with welding are not limited to offshore structures and large vessels only but it is common in small crafts especially high-speed crafts as well.

Classification societies play an important role in recommending and requiring different welding configurations (DC, CI and SI) based on numerous design aspects. In so far, classification requirements are concerned, DC welding is required and recommended in the bottom shell structure in the impact area of high-speed craft [1] and often it is a point of discussion in the small and high-speed craft industry.

The point of discussion is mainly due to numerous problems associated with double continuous welding (DC) in small craft. In general, DC welding is not preferred and sometimes may not be even feasible and practicable due space constraints and excessive distortion.

1.1 MOTIVATION

Currently, several non-displacement and semi-displacement crafts are used for crew transfer, coastal patrol, rescue operation, and various other naval operations. These crafts are inherently high-speed crafts. Due to the nature of its operation, hull geometry, and mainly velocity; high-speed craft encounters high hydrodynamic wave loads, slamming and substantially high bottom impact pressure, which mainly arises due to hull and wave interaction and motion of the craft itself.

Welding is primary joining method used to join adjacent structures in shipbuilding process especially where aluminium alloys or steel are used as a construction materials. Welding methods can be of various types based on load application (Impact, slamming, surge, hydrostatic, and hydrodynamic). DC welding is the most common approach in shipbuilding industry despite creating bigger heat affected zone (HAZ), structural distortion, usage of a substantial amount of welding consumables, man-hours and enhanced project completion time. Optimisation of welding requirements is need of the time.

Recommendation to use DC welding is based on historical approach, and there are potential opportunities to optimise and generalise the welding requirements based on direct calculation for structural behaviours and model test to ascertain motion characteristics and seakeeping parameters accurately.

Finding suitable alternatives to DC welding may benefit shipbuilding industry in terms of:

- Reduction in distortion and HAZ;
- Less complex welding sequence;
- Labour cost;
- Material cost;
- Project completion time.

1.2 OUTLINE OF THE PAPER

It is complex process to identify and generalise craft's impact area and find suitable welding methods for all range of mono-hull craft which is dependent on various factors but not limited to hull geometry, displacement, and floating position, craft speed, range of refuge, operational area and environment [1]. For this study, a high-speed craft is selected. Table 1.1 shows craft's main particulars.

Table 1.1 Main particulars of the high-speed craft

Main particulars	Unit	Dimension
Waterline length	m	25.25
Rule length (L_R)	m	24.50
Depth	m	3.64
Breadth	m	6.60
Summer Draught	m	1.44
Maximum Speed	Knots	25
Moulded Displacement	Tonnes	95.0
Significant Wave Height	m	2.0
Taylor Quotient		4.98
Service Area		G3
Service Type		Patrol

Parameters like design panel pressure, which is mainly dependent on hydrostatic, hydrodynamic wave pressure, pitching pressure, bottom impact pressure and vertical acceleration, along with identification of slamming zone are evaluated in accordance with LR SSC Rules [2]. Estimated panel pressure is applied on hull structures and structural evaluations in terms of strength and fatigue behaviours are carried out. Figure 1.1 shows outline of the study.

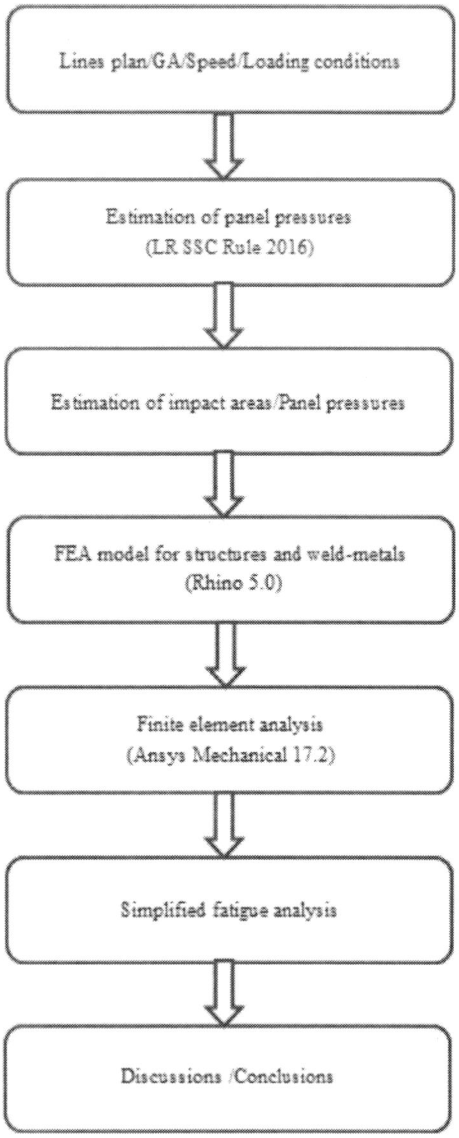

Figure 1.1 Outline of the Study

1.2.1 Limitation of the Study

High-speed craft usually experiences significantly high pressure throughout the length of the craft. Extreme pressure is observed in length between midship to $0.75L_R$ from aft peak and study is limited for this area. High panel pressure is mainly observed due to continuous impact pressure, which is primarily caused due to craft vertical acceleration. Figure 3.3 shows the applied panel pressures along the length of the craft.

The study is limited to different welding configurations used for joining secondary stiffeners to shell plate and adjoining primary members. The connection of primary members with shell plate is considered to be of DC welding and the same assumption is applied throughout.

2. LITERATURE REVIEW

Comparison of strength characteristics among different welding configurations (DC, CI and SI) for marine application is limited to steel as a construction material. Whereas, aluminium alloy as a construction material is concerned, study related to strength and fatigue behaviours in association with different welding configurations is very limited.

Miller et al. [3] carried out extensive experimental research to ascertain the root causes behind the catastrophic failure of two inland tank barges (Buffalo 286 and Buffalo 292).
Collapse occurred in calm weather and considered mainly due to buckling failure of the deck. These barges were of steel construction and decks were constructed using intermittently welded serrated longitudinal deck stiffeners which are rarely in use now. Research was focussed on the mechanism of collapse in a special type of non-continuously welded stiffened plates, namely intermittently welded serrated stiffened plates, which were applied in the construction of these inland tank barges. Research concludes that:

Serrated stiffeners reduce panel strength. The average value for the ultimate compressive load of the panels using intermittently welded serrated stiffeners were less than that of intermittent and continuously welded stiffeners. Continuously welded stiffeners make the strongest panels. The average ultimate load carried by the continuously welded panels was 5% greater than the average ultimate load carried by the intermittently welded panels and 9.2% higher than average ultimate load carried by the serrated intermittently welded panels. Further, it was also concluded that intermittent fillet welding is most common welding technique and provide 5% more ultimate strength than intermittently welded serrated panels and 5% less than that of continuously welded panels and simultaneously require approximately 80% less length of the weld.

Khedmati et al. [4] carried out a comparative study for strength behaviours and ductility characteristics in association with three different welding construction methods of stiffened plates. The study is done for steel ship structures in main deck area. Applied loads were primarily in-plane axial compression. Study was carried out for different weld configurations (DC, CI and SI). Study concludes in comparison with continuously welded stiffened plates; the ultimate strength is much more reduced in case of chain intermittent fillet welds than the case of staggered intermittent fillet welds.

3. BOTTOM PRESSURE ESTIMATION

This section lists methods and steps to estimate the expected panel pressure in the bottom shell plate. It is evaluated based on the LR SSC Rules [2].

Figure 3.1 [5] illustrates numerous types of loads encountered by a high-speed craft. Loads are mainly caused due to motions of the craft, static and dynamic sea pressure, vibratory nature from propulsive machinery, cargo, weather, hull girder and equipment loads.

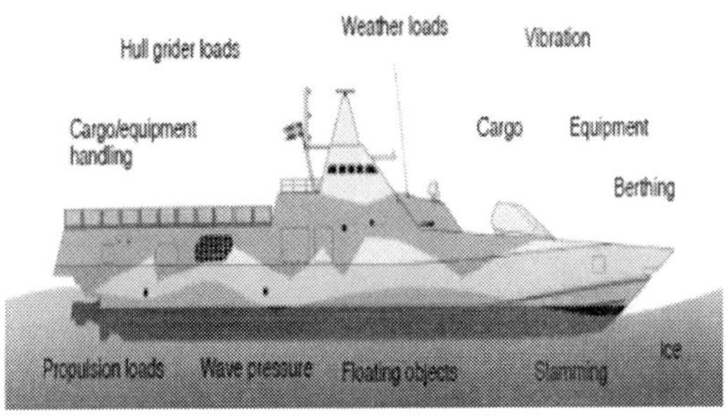

Figure 3.1 Nature of loads

High-speed craft is also termed as non-displacement or planning craft. High-speed craft, during its operation, is capable of attaining speed, equal or greater than volumetric speed number (F_v). In addition; operate at Taylor Quotient (Γ) greater than 3.0. Definition of Volumetric speed number (Fv) and Taylor Quotient (Γ) is shown below.

$$F_v = 7.19\,\nabla^{1/6}$$
$$\Gamma = V_s / \sqrt{L_{WL}}$$

On another hand, craft with Taylor Quotient (Γ) less than 3.0 are considered as displacement or semi-displacement craft. Other significant difference between displacement and non-displacement craft is the application of panel pressure at the bottom and side shell. Bottom and side shell of displacement craft are mainly subjected to combined load of hydrostatic and hydrodynamic loads. Whereas, the bottom and side shell of non-displacement craft are subjected to the combined load of hydrostatic and hydrodynamic loads or bottom impact pressure, whichever is higher. Figure 3.2 [2] shows the application of hydrostatic and hydrodynamic pressures on the bottom shell plating.

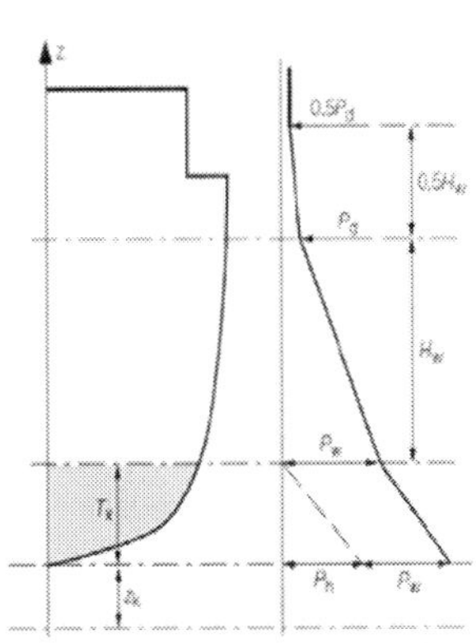

Figure 3.2 Hydrostatic and hydrodynamic pressures on the shell plating

Bottom impact pressure is a dynamic pressure and predominant in non-displacement craft and often exceeds combined load of hydrostatic and hydrodynamic loads. Bottom impact pressure is directly proportional to vertical acceleration at longitudinal centre of gravity (L_{CG}), displacement and significantly affected by hull form of the craft. Further, vertical acceleration at L_{CG} is dependent on Taylor Quotient, significant wave height, breadth of the hull between the chines, dead-rise angle at the L_{CG} and running trim angle [2].

Accordingly, bottom impact pressure on the bottom and side shell for non-displacement craft, a predominant dynamic load, is directly proportional to the speed of the craft, displacement, significant wave height and breadth of the hull between the chines. Due to its dynamic nature of the predominant load, fatigue life of non-displacement craft mainly depends on bottom impact pressure.

High-speed craft is significantly shorter than most displacement crafts, operate in relatively protected waters and operate at higher forward speeds. The effect of these differences needs to be addressed when establishing the loading history for high-speed craft. There are two factors that will help to reduce the relative magnitudes of the primary hull girder bending moments and stresses acting throughout a high-speed craft relative to a displacement hull, i.e., length and sea-state. The structural design of many shorter crafts, such as high-speed craft, is governed by the secondary loads, mainly caused due to hydrostatic, hydrodynamic and bottom impact pressure. It is not necessary to increase scantlings which are based on secondary loads to account for the primary loads and stresses due to global bending. This trend is further emphasised for a high-speed craft that operates in protected waters and will not experience the same extreme wave heights and sea-states that generate the higher bending moments in shorter craft [6].

These secondary loads are mainly caused due to hydrostatic, hydrodynamic and bottom impact pressure can be estimated based on different Classification Societies rules. During the study, various components of panel pressures, i.e. hydrostatic pressure, impact pressure, fore-body impact pressure and hydrodynamic wave pressure are estimated, based on Part 5 of Lloyd's Register Rules and Regulations for the Classification of Special Service Craft, July 2016 [2].

3.1 LR SSC RULES FOR PANEL PRESSURES

The panel pressures estimated based on LR SSC Rule are derived based on experimental and theoretical studies complemented with the service experiences. The loads imposed by the environment are based on the environmental condition in which, a craft is designed to operate. Pressure arises as the craft advances in a seaway and is loaded and stressed randomly by dynamic forces and moments. LR offers various geographical limits based on the range of refuge for the craft and consequently recommend expected significant wave height, craft may encounter throughout her operations. Table 3.1 [2] shows the relation between operational areas and operating significant wave height.

Table 3.1 Minimum operating significant wave height (m)

Service group	Minimum significant wave height (m)	Range of refuge (Nautical miles)
G1	0.6	-
G2	1.0	-
G2A	1.5	-
G3	2.0	150
G4	4.0	250
G5	4.0	350
G6	4.0	Unrestricted

Craft considered for this study is operating in the area "G3" as defined in the LR SSC Rules, is exposed to a significant wave height of 2.0m. Considering various parameters discussed in Part 5 of LR SSC Rules; Figure 3.3 [2] shows estimated hydrostatic, hydrodynamic and bottom impact pressures.

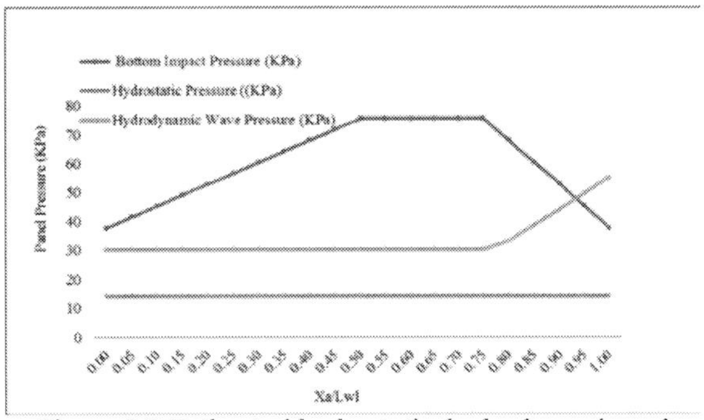

Figure 3.3 Estimated hydrostatic, hydrodynamic and bottom impact pressures along length of the craft

4. STRENGTH EVALUATION

This section covers considerations and methods for preparation of models and analysing the structures. Solid geometry model of whole structures including weld-metals is prepared using commercial tool Rhino 5.0 from Robert McNeel & Associates and structural finite element analysis is done with the help of Ansys Mechanical 17.2 from Ansys Inc.

Steel and aluminium are primary construction materials used in the shipbuilding industry. Aluminium is most common material used for building high-speed crafts. Boat and the high-speed catamaran builders, for the first time during 1990's, used aluminium panels for marine applications [7]. The main roles of these panels are strengthening against in-plane compression and withstand lateral pressure.

Whereas aluminium as hull construction material is concerned, it has unique properties in comparison to other hull construction materials like steel. It is well known; aluminium weldments reduce the strength of parent metal through HAZ and strength of welded joint cannot be utilised for un-welded strength. For some aluminium alloys, strength in HAZ is reduced to almost 50% of that of the un-welded aluminium alloy. Table 4.1 [2] shows the proof stress of aluminium alloys in welded and un-welded conditions.

Table 4.1 Proof stress of aluminium alloys in welded and un-welded conditions

Alloys	Condition		0.2% Proof stress MPa Un-welded	0.2% Proof stress MPa Welded
5083	O/H111		125	125
6061	T5/T6	Rolled	240	125
		Extruded: Profile	240	125

The specific density of aluminium is equal to one-third that of steel. In terms of corrosion resistance, the exposed surface of aluminium combines with oxygen to form a thin inert aluminium oxide film which blocks further oxidation. Contrary, steel must be always corrosion protected in any kind of environment. The structures made of aluminium alloys are more sensitive to thermal variations because the coefficient of thermal expansion of aluminium is twice that one of steel. Contrary, residual stresses produced by constraining thermal deformations are about 30 percent lower than those in steel structures because they are proportional to the product of young modulus and thermal elongation coefficient [8]. From fatigue point of view, aluminium does not reflect similar behaviour as steel regarding endurance limit. Steel may withstand a theoretically infinite number of the cycles for some stress range. However, aluminium generally does not reflect well-defined endurance limit and tend to fail regardless of stress range values. Further; having a different constitutive stress-strain relationship from that of structural steel, it is impossible to extend the formulations used for strength of steel panels to the aluminium panels [9].

4.1 FINITE ELEMENT ANLYSIS GEOMETRIES

Solid geometries of structures including weld shape is modelled using Rhino 5.0, a commercial 3-D computer graphics and computer-aided design (CAD) application software developed by Robert McNeel & Associates. Rhino geometry is based on the NURBS mathematical model.

Structures considered for analysis is located between midship and $0.75L_R$ from aft peak, where craft encounters maximum dynamic impact pressure. Structures have been modelled for five frames in the longitudinal direction and along the transverse direction; geometry extends from centreline girder to side shell chine to mainly simulate the deflection and structural behaviour accurately.

Scantlings in the considered area consist of shell plate of thickness 6.0mm and are of aluminium 5083-OH11 grade. Stiffeners are 80x6 flat bar of aluminium 6061-T6 grade. Bottom shell of the craft is longitudinally framed and stiffeners are spaced at an equidistance of 235 mm. Primary transverse members are 215x6 + 50x6 T section and are of grade aluminium 6061-T6. These are spaced at a frame spacing of 650 mm. Table 4.1 shows the mechanical properties of these alloys.

Figure 4.1 and Table 4.2 show recommended weld's throat thickness, weld length and spacing. These are estimated based on LR SSC Rules [2].

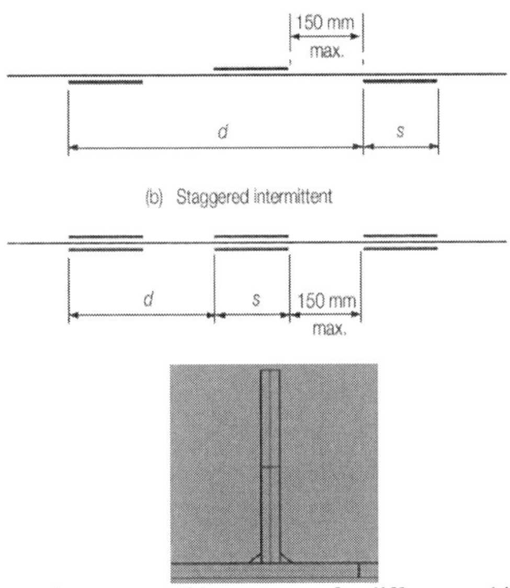

Figure 4.1 Weld parameters for different weld configurations and typical weld profile

Table 4.2 Weld parameters for different weld configurations (DC, SI and CI welding)

Welding configurations	Weld length	Weld spacing	Throat thickness
	s	d	T_t
	mm	mm	mm
DC	-	-	3.0
SI	65	260.0	4.5
CI	65	195.0	4.5

Three different models, each for DC, SI and CI welding configurations is prepared in accordance with weld parameters and scantlings indicated in subsection 4.1. Figure 4.2 shows plan view of modelled structures in DC, CI and SI welding configurations.

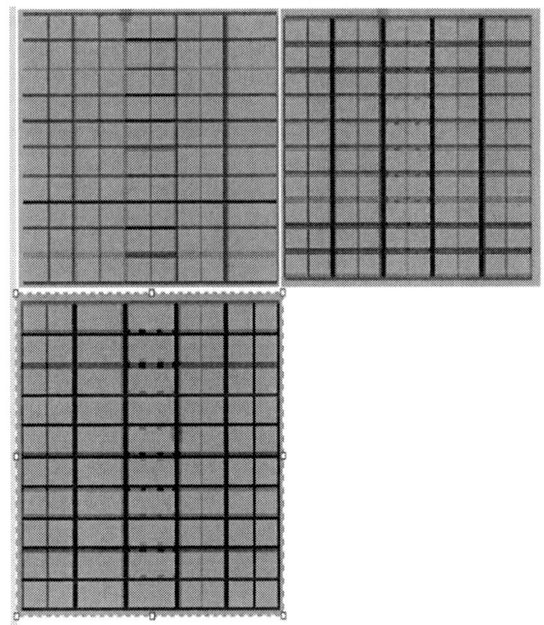

Figure 4.2 Plan view of modelled structures in DC, SI and CI welding configurations.

4.2 FINITE ELEMENT ANLYSIS

Finite element analysis (FEA) is carried out with Ansys Mechanical 17.2, a commercial FEA tool developed by Ansys Inc.

Structures have been modelled for five frames in the longitudinal direction. However, while considering and analysing the results, mainly weld geometries, longitudinal stiffeners and plate assembly between adjacent transverse members are considered.

Stiffeners are connected with the plate and adjoining primary transverse members with solid welds geometries. For DC welding configuration; contact between stiffeners to plate and stiffener to adjoining primary transverse members is considered to be of frictional contact with the friction coefficient of 0.2. Whereas, CI and SI welding configurations are concerned; contact between stiffeners to plate and stiffener to adjoining primary transverse members is considered to be of frictional contact with the friction coefficient of 0.1. Boundary conditions are considered to be fixed and all six degrees of freedom (U, V, W, θ_x, θ_y, θ_z) is restrained at side shell chine, centreline girder and forward and the aft end of modelled geometry. The tetrahedron and hexahedron solid element, among the element library of Ansys, is selected for meshing the stiffened plates and weld geometries.

4.2.1 Convergence

The accuracy of response generally depends on many factors but not limited to the mesh size of elements. Finite element mesh is used to sub-divide the model into small elements over which equations are solved and the response is obtained. Mesh refinement is a key factor to get the results accurately.

There are few ways to establish appropriate mesh size for structures. Generally, mesh refinement throughout the whole structures is done by reducing the mesh size globally. However, in some cases, it may not be a suitable option. Some of the commercial FE software offers adaptive mesh refinement tools. It uses error estimation strategy and calculates the error throughout structures and location where the error is extreme; it further refines mesh size in that area.

For this case, converging mesh size is estimated for DC welding configuration only. Element size is varied from 24 mm to 12 mm and associated equivalent stress is estimated and plotted. Figure 4.3 shows equivalent stress contour for an element size of 12 mm, and Figure 4.4 shows equivalent stress (MPa) vs. mesh element size (mm).

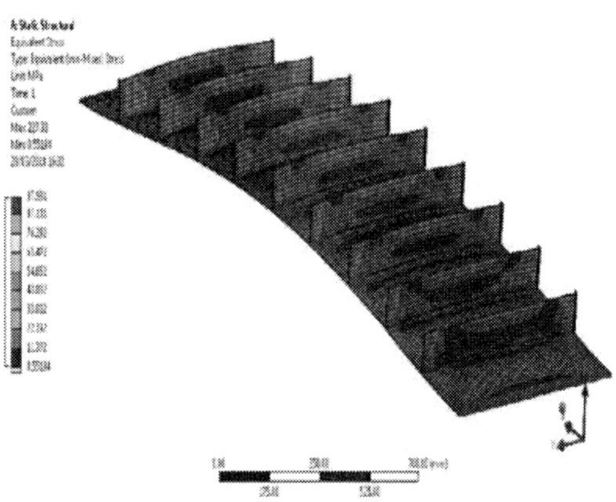

Figure 4.3 Equivalent stress contour for an element size of 12 mm

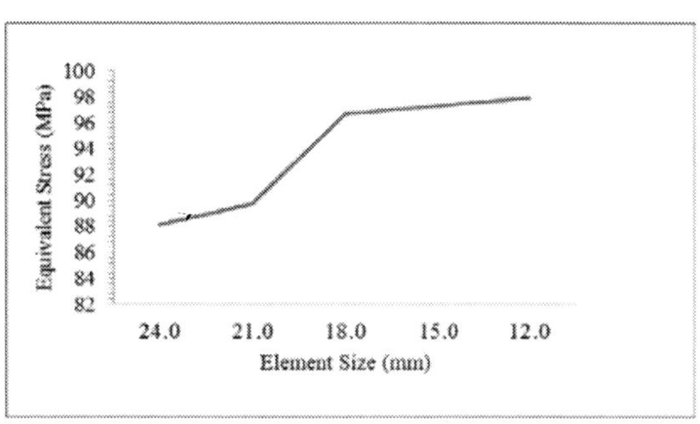

Figure 4.4 Equivalent stress (MPa) vs. mesh element size (mm)

Based on convergence plot; selected element size of 12 mm as shown in Figure 4.4 and other conditions as described in subsection 4.2; repetitive studies have been carried out for different weld configurations (DC, SI and CI welding) to evaluate equivalent stress and shear stress for whole structures including weld-metals and weld-metals exclusively.

Figures 4.5 and 4.6 show the contours of equivalent and shear stress for DC, SI and CI welding configurations respectively. These also show the equivalent and shear stress contours in whole structures including weld-metals and weld-metals exclusively.

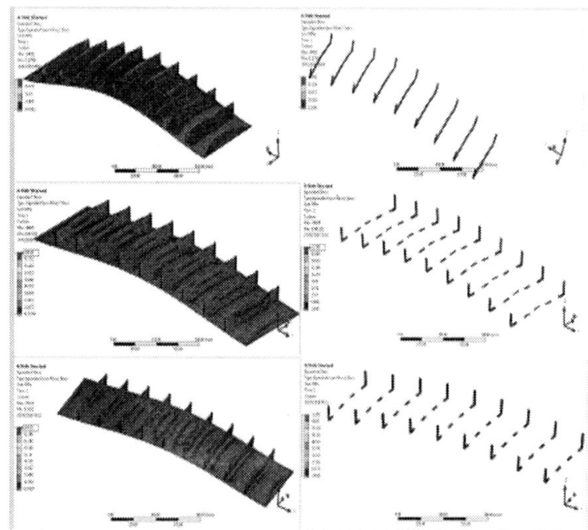

Figure 4.5 Equivalent stress contours in DC, SI and CI welding configurations

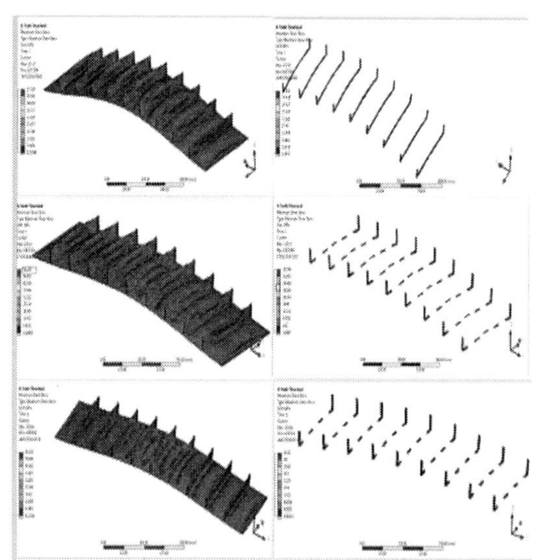

Figure 4.6 Shear stress contours in DC, SI and CI welding configurations

In summary, Figures 4.7 and 4.8 show equivalent and shear stress respectively in whole structures including weld-metals and weld-metals exclusively in different weld configurations.

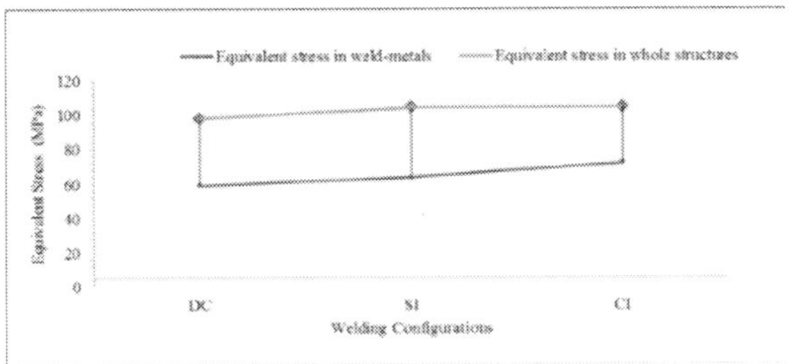

Figure 4.7 Equivalent stress in whole structures including weld-metals and weld-metals exclusively in different weld configurations.

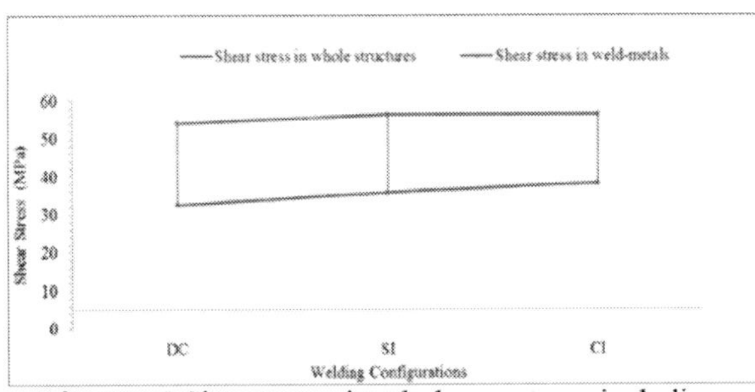

Figure 4.8 Shear stress in whole structures including weld-metals and weld-metals exclusively in different weld configurations.

5. FATIGUE LIFE EVALUATION

This section covers methods and considerations to estimate fatigue design life. For evaluation of fatigue life, simplified fatigue approach is used. S-N curves for this study are based on EuroCode 9, Part 2: Structures Susceptible to Fatigue; Design of Aluminium Structures [10].

The main factors that contribute to general fatigue problems are the material used for construction, welding methods, and connection details, along with the stress range and the number of stress cycles. High-speed crafts are usually designed for optimized structural weight, which results in increased flexibility of the hull. The combination of the increased hull flexibility and the higher encounter frequency due to high operational speed would tend to accelerate fatigue damage [11].

5.1 METHODOLOGY FOR EVALUATION

Generally, high-speed crafts operate close to their structural design limits and fatigue is a more pronounced design factor. The accumulation of fatigue damage due to cyclic loading must be considered in the design of craft's structural components. The principal source of cyclic stresses in the structure will be motion and wave-induced loads occurring at high encounter frequencies due to the high forward speeds. Figure 5.1 shows the flow diagram for fatigue life calculation.

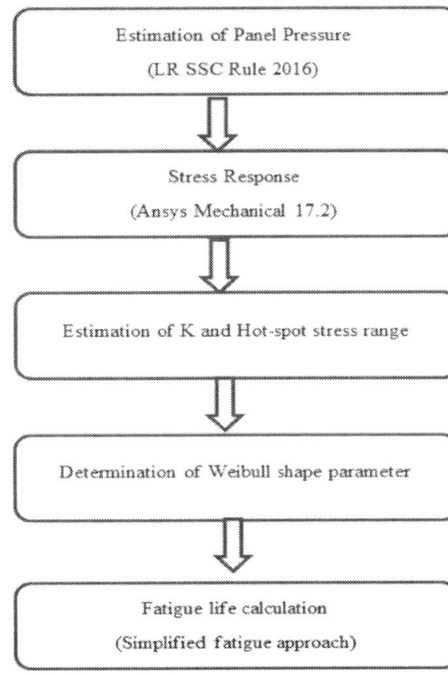

Figure 5.1 Flow diagram for fatigue life calculation

Nominal stresses are the most fundamental stresses calculated for every design. These are the global field stresses found in a structure using classical elastic stress analysis techniques without the impact of any stress concentrations, discontinuities or other geometric amplification effects. They can be determined through simple hand calculations such as axial stress or bending. For more complex structures they can be calculated with finite element tools.

Nominal stress range is estimated based on load due bottom impact pressure and hydrostatic load. Figure 3.3 shows dynamic impact pressure and hydrostatic pressure along the length of the craft. Nominal stress range is estimated based on maximum principal stress and mainly at weld-toe. Figure 5.2 shows the contour of maximum principal stress for DC, SI and CI welding configurations.

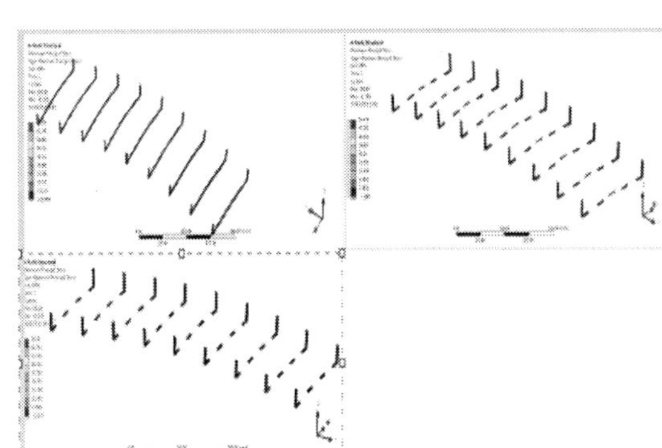

Figure 5.2 Maximum principal stress at weld-toe for DC, SI and CI welding configurations

DNV [12] offers guidelines about the interrelation between hot-spot stresses and nominal stresses. Fatigue life is influenced significantly by hot-spot stresses, which are mainly caused due to stress concentration, residual stress, weld imperfection and structural mismatch of geometry. The hot-spot stress is obtained by multiplication of nominal stress by K-factor and can be defined as:

$$K = \sigma_{hot\text{-}spot}/\sigma_{nominal}$$

Accordingly, wherever structures are of a welded connection, the relationship between hot-spot stress range and nominal stress range can be defined as

$$\Delta\sigma_{hot\text{-}spot} = K.\Delta\sigma_{nominal}$$

All stress enhancers are to be considered when evaluating the hot-spot stress range. This can be estimated by multiplication of K-factors arising from different causes. The resulting K-factor to be used for calculation of hot-spot stress and is derived as:

$$K = K_g . K_{te} . K_{ta} . K_n$$

K_g = Stress concentration factor due to the gross geometry of the detail considered;
K_{te} = Additional stress concentration factor due to eccentricity tolerance (typically used for plate butt weld connections only);
K_{ta}= Additional stress concentration factor due to angular mismatch (typically used for plate butt weld connections only);
K_n = Additional stress concentration factor for un-symmetrical stiffeners on laterally loaded panels, applicable when the nominal stress is derived from simple beam analysis.

The K-factors in way of welds depends on the level of workmanship. From a design point of view, it is obviously important to keep K-factors as low as possible (with a target value of 1.0). Stress concentration factor, which significantly affects fatigue life, depends upon ratio ρ/T (ratio of weld-toe radius to plate thickness). It is observed that the biggest increase of stress concentration factor occurs when ρ/T is less than 0.2. Further, ratio of weld throat thickness to plate thickness ratio also affects stress concentration factor. The increase of fatigue strength of welded joints produced by weld geometry modification methods could be attributed to the increase of weld-toe radius [13]. Figure 5.3 [13] shows influence of weld throat thickness on stress concentration factor.

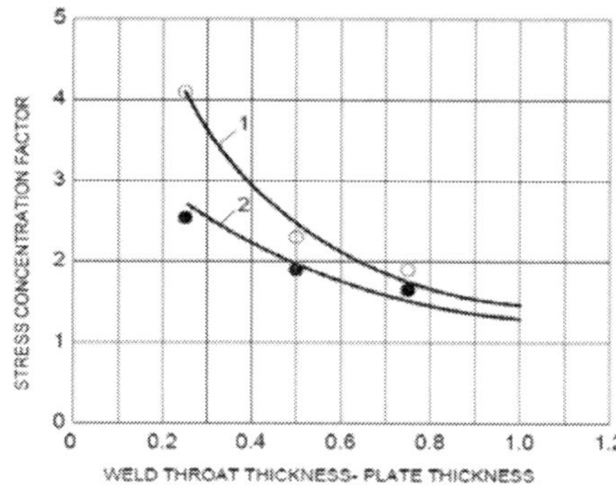

Figure 5.3 Influence of weld throat thickness on stress concentration factor

This concentration factor will depend on the toe angle (μ) and particularly the ratio ρ/T. Low values of μ and high values of ρ/T are favourable [14]. If found unfavourable, fatigue life, initiation and propagation parts of the life are reduced. The problem is due to the fact that when the plate thickness is increased, the variable toe radius is not controlled. The toe radius is related to the geometry of the weld pass, close to fusion line. This radius is likely to have the same value regardless of the chosen plate thickness. Hence, the ρ/T ratio will inevitably decrease when T increases [14].

Stress concentration for fillet weld in association with toe angle (μ) and particularly the ratio ρ/T is [14] [15]:

$$K_g = 1+0.51.(\mu)^{0.25}.(T/\rho)^{0.47}$$

K_g can more easily be revealed by an FEA of the detail. This type of analysis will, however, require special qualities of the model regarding volumetric element type and mesh. However; for this study, K_g is estimated based on empirical formulae [14] [15] based on weld geometry. Toe angle (μ) is considered to be 30 degrees (0.52 radian) and fictitious toe radius of 1.0mm. Fictitious notch radius of 1.0 mm can be applied to aluminium welded joints from plates with the thickness between 5 to 25mm [16].

Additional stress concentration factor (K_{ta}) due to angular mismatch is estimated in accordance with DNV recommendation for Fatigue assessment of ship structures [12]. Empirical formulae to estimate additional stress concentration factor (K_{ta}) is shown below.

$$K_{ta} = 1+(\lambda/4).\upsilon.P_b/T$$

Angular mismatch (υ) is estimated to be based on assumed plate deflection of 2.0 mm for the considered plate thickness of 6.0 mm. Plate width for considered configuration is 235 mm. The coefficient for end supports (λ) is recommended to be 3.0 [12].

Other parameters like additional stress concentration factor due to eccentricity tolerance (K_{te}) and additional stress concentration factor for un-symmetrical stiffeners on laterally loaded panels (K_n), which may impact hot-spot stress range is considered to be 1.0 and cautions are to be exercised in case of otherwise and further studies are to be carried out.

5.2 SIMPLIFIED FATIGUE APPROACH

This section provides information on assessment of design life based on simplified fatigue approach for application to aluminium weldments in the high-speed craft. The approach is based on simplified fatigue assessment method, which is detailed in section 5 of Guide for Fatigue Assessment of Offshore Structures [17]. The simplified fatigue assessment method employs the Weibull distribution to model the long-term distribution of sea states and Miner's cumulative damage rule. Figure 5.4 [18] shows Miner's linear cumulative damage rule for estimating fatigue life under variable amplitude loading.

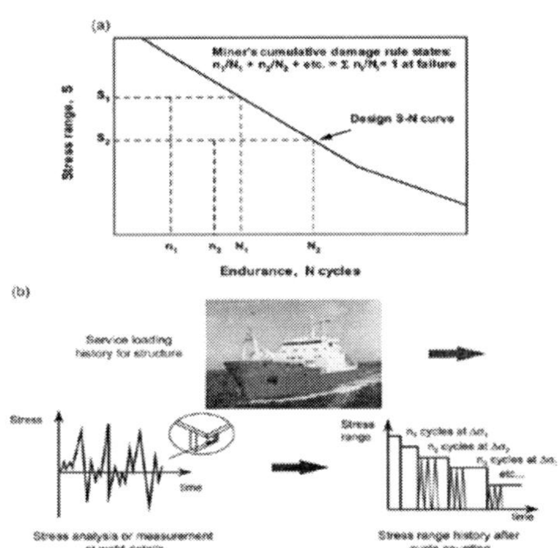

Figure 5.4 Miner's linear cumulative damage rule for estimating fatigue life under variable amplitude loading; (b) analysis of fatigue loading for cumulative damage calculations

Weibull probability distribution, which is very common in marine and offshore application, is used. Based on this approach, estimated Weibull distribution, expressions for fatigue damage at the design life (N_T) of the structure are derived. Weibull shape parameter is the most important parameter, which varies between 0.7 to 1.4, depending on the dominant period of the structural response and the considered wave environments [17]. For high-speed craft, there is no relevant information is available to estimate Weibull shape parameter. Accordingly, Weibull shape parameter is estimated based on ships parameters as detailed in ABS Rules [19]. Based on recommended value $\alpha = 1$ for bottom structures, including inner bottom, and side shell and longitudinal bulkhead structures within 10% of depth from the bottom [19]; estimated Weibull shape parameter (Y) for the craft is 1.05 and in accordance with following empirical formulae [19].

$$Y = 1.40 - 0.2.\alpha.L_R^{0.2}$$

In the simplified fatigue assessment method, the two-parameter; Weibull shape and scale parameter are used to model the long-term distribution of fatigue stresses. The cumulative distribution function of the stress range can be expressed as:

$$F_s(S) = 1 - \exp(-(S/\delta))^{Y} \qquad S > 0$$

Based on the long-term distribution of stress range, a closed form expression for fatigue damage can be derived. The major feature of the simplified method is that appropriate application of experience data can be made to establish or estimate the Weibull shape parameter, thus avoiding a lengthy spectral analysis.
The other major assumptions underlying the simplified approach are that the linear cumulative damage (Palmgren-Miner) rule applies, and that fatigue strength is defined by the S-N curves [17].

The cumulative fatigue damage for the two-segment case of S-N curve is expressed as:

$$D = (N_{T.}\ \delta^{m}/A).\ \Gamma\ (m/Y+1, Z) + (N_{T.}\ \delta^{r}/C).\ \Gamma_0(r/Y+1, Z)$$

$\Gamma(a,z)$ and $\Gamma 0(a,z)$ are incomplete gamma functions (integrals z to ∞ and 0 to z, respectively).

$$\Gamma(a,Z) = \int_Z^{\infty} t^{a-1}e^{-t}dt = \Gamma(a) - \Gamma_0(a,Z)$$

$$\Gamma_0(a,z) = \int_0^{Z} t^{a-1}e^{-t}dt$$

$$a = 1 + m/Y$$

$$b = 1 + r/Y$$

$$Z = (S_Q/\delta)^{Y}$$

$$\delta = S_R/(\ln N_R)^{1/Y}$$

In summary, in association with estimated K-factors; Table 5.1 and Figure 5.5 show estimated nominal stress range and hot-spot stress range at weld-toe.

Table 5.1 Nominal stress range and hot-spot stress range at weld-toe

Welding Methods	Nominal stress range at weld-toe	K_g	K_{ta}	Hot-spot stress range at weld-toe
	MPa			MPa
DC	49.91	2.01	1.5	150.5
SI	54.35	2.01	1.5	163.9
CI	56.28	2.01	1.5	169.7

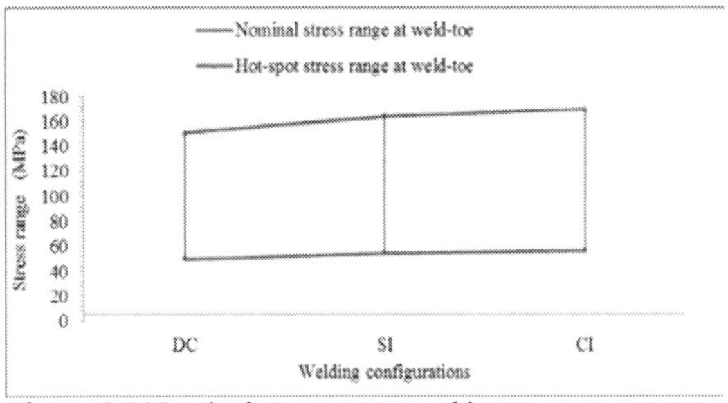

Figure 5.5 Nominal stress range and hot-spot stress range at weld-toe

5.3 SELECTION OF S-N CURVES

Selection of S-N curves is other important parameter to be considered with great precision to estimate the fatigue life reasonably. Especially, wherever, aluminium alloys are used, cautions are to be exercised to find suitable S-N curves for structures and loading patterns. For this study, S-N curves selected based on EuroCode 9 and are indicated in Figure 5.6 [10] and associated numerical values are tabulated in Table 5.2 [10]. These curves are recommended to be suitable for longitudinally welded attachments and can be used for various welding methodologies like DC, CI and SI welding configurations. As recommended, S-N curve 40-4.3 is considered for DC welding configuration and curve 35-4.3 is considered for CI and SI welding configurations.

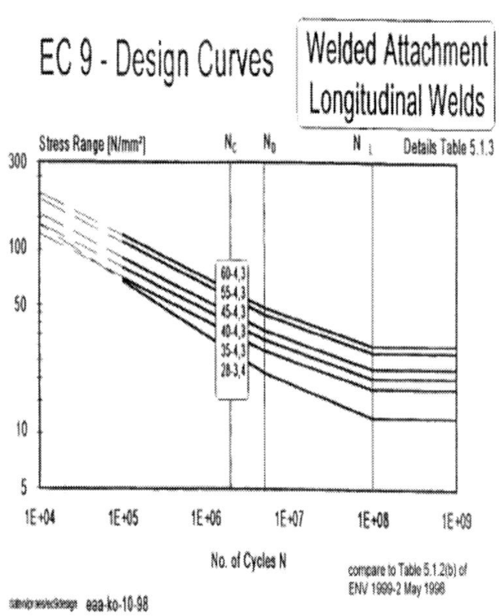

Figure 5.6 S-N Curves for members with welded attachments

Table 5.2 Amended numerical values for selected S-N curves for members with welded attachments

Details Category	A	m	C	r	N_q	S_q
	Cycles		Cycles		Cycles	MPa
60	8.86E+13	4.3	2.07E+17	6	5.00E+06	48.5
55	6.08E+13	4.3	1.20E+17	6	5.00E+06	44.4
45	2.58E+13	4.3	3.40E+16	6	5.00E+06	36.4
40	1.55E+13	4.3	1.62E+16	6	5.00E+06	32.3
35	8.70E+12	4.3	6.78E+15	6	5.00E+06	28.3
28	4.00E+12	4.3	7.35E+14	6	5.00E+06	21.4

5.4 ESTIMATION OF DESIGN CYCLES

Design cycles for the craft are estimated based on assumption that craft will operate eight hours per day and moves at speed of 25 and 20 knots throughout her design life of 20 years. The angle of attack between ship and wave (α) is 45, 90,135 and 180 degrees and distributed equally throughout her design life. Craft encounters maximum frequency in head sea condition. Numerically, encounter frequency is estimated based on the following equation.

$$\omega_e = \omega.(1- ((\omega.\ V_s/g).\text{Cos}\ \alpha)))$$

Based on assumption stated above, during design life of 20 years, operating eight hours per day, craft expect to encounter 27.8 million cycles.

5.5 ESTIMATION OF FATIGUE DAMAGE

Simplified fatigue approach; using equations discussed in subsection 5.2 are used to estimate fatigue damage during design life of the craft. Hot-spot stress range at weld-toe as detailed Table 5.1, S-N curves parameters detailed in Table 5.2 are used for estimation and shown in Table 5.3 as input parameters for different weld configurations.

Table 5.3 Fatigue input parameters (DC, SI and CI welding)

Welding Methods	Class	m	A	r	C	N_R	S_Q
			Cycles		Cycles	Cycles	MPa
DC	40	4.3	1.55E+13	6.3	1.62E+16	2.78E+07	32.3
SI	35	4.3	8.70E+12	6.3	6.78E+15	2.78E+07	28.3
CI	35	4.3	8.70E+12	6.3	6.78E+15	2.78E+07	28.3

In summary, Table 5.4 and Figure 5.7 show estimated fatigue life in cycles and years.

Table 5.4 Design life at weld-toe (DC, SI and CI welding)

Welding Methods	δ	1+m/Y	1+r/Y	z	Γ ((1+m/y,z)	Γ0 (1+r/y,z)	Fatigue damage	Estimated fatigue life	Estimated fatigue life
								Cycles	Years
DC	9.97	5.11	7.02	3.43	21.31	43.67	0.9	3.09E+07	22.3
SI	10.85	5.11	7.02	2.73	24.53	15.77	2.4	1.14E+07	8.2
CI	11.24	5.11	7.02	2.63	24.91	13.24	2.9	9.75E+06	7.0

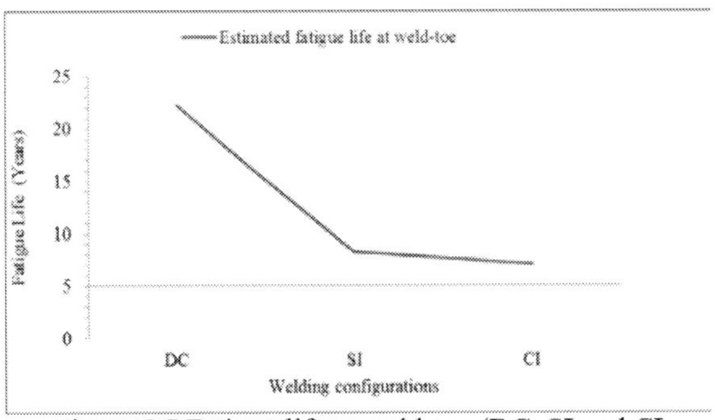

Figure 5.7 Fatigue life at weld-toe (DC, SI and CI welding)

6. CONCLUSIONS

Impact pressure on the bottom and side shell for non-displacement craft, a predominant dynamic pressure, is directly proportional to speed of the craft. The In-plane, compressive loads due global bending of the craft are considered to be insignificant due inherent size of craft to be small in the length. Accordingly, the craft is expected to encounter bottom impact pressure of 75.35 KPa between midship and $0.75L_R$ from aft peak and will govern the design criteria for the craft.

Equivalent stress in whole structures including weld-metals is 6.5%, higher in CI and SI welding configurations than that of in DC welding configuration. Equivalent stress in weld-metals exclusively is 7.8%, 21.5% higher, in SI and CI welding configurations respectively, than that of in DC welding configuration.

Shear stress in whole structures including weld-metals is 4.1%, higher in CI and SI welding configurations than that of in DC welding configuration. Shear stress in weld-metals exclusively is 9.8%, 17.2% higher, in SI and CI respectively, than that of in DC welding configuration.

Based on selected S-N curves and K-factors as considered in the study, Fatigue life in CI and SI welding configurations is significantly reduced and observed to be almost one-third of that observed in DC welding configuration.

7. RECOMMENDATIONS FOR FUTURE WORKS

During this study, following areas require further improvements and research works:

Estimated panel pressure is based on the numerical and empirical basis and further model test may be required to validate these results.

This study doesn't account in-plane compressive load and assumed comparatively insignificant due to the length of the craft. However, this assumption is to be validated or in-plane compressive load are to be incorporated in further evaluations.

Weibull shape parameter is the most important factor to estimate fatigue life. For this study, it is estimated based on ship's parameters as detailed in ABS Rules [19]. However, ABS Rules [19] is suitable for large vessels of length more than 150 meters and may not be suitable for high-speed mono-hull craft. Further studies are to be carried out to estimate Weibull shape parameter accurately.

For this study, residual stresses and implications due to start and stop in intermittent welding process are not considered and are to be considered in further evaluation.

Study is also limited to scantlings between midship to $0.75L_R$ from aft peak. Further studies are to be carried out throughout the length of the craft in association with different scantlings configurations to understand and generalise fatigue behaviours.

Some of the considerations made to estimate K-Factors are based on steel its suitability for usage of aluminium is to be studied further.

Ratio of weld throat thickness to plate thickness also affects stress concentration factor and is to be studied further.

8. ACKNOWLEDGEMENTS

Express profound gratitude to Dr Arun Kr Dev for encouragement and guidance throughout the postgraduate study. Also indebted to Lloyd's Register (LR) for immense support and assistance.

Opinions expressed in this paper are exclusively those of the authors.

9. AUTHORS BIOGRAPHY

Mr **Vimal Kumar** holds the position of Plan approval specialist at Lloyd's Register. He is responsible for plan appraisal of special service crafts and various types of ships. He gained MSc in Marine Technology from Newcastle University, UK.

Dr Burak Can Cerik gained a PhD from University of Ulsan, Korea, lectured at Newcastle University, School of Marine Science and Technology for three years, before returning to Korea to a researcher position in Inha University. He specialises in the analysis of marine structures against extreme loadings.

10. REFERENCES

[1] Lloyd's Register, Guidance Note – Extent of double continuous welding for special service craft

[2] Lloyd's Register, Rules and Regulations for the Classification of Special Service Craft, July 2016.

[3] M. S. Miller, J. P. Nadeau and G. J. White, "Longitudinal Stiffened Panels-A Comparative Analysis of the Compressive Strength of three common Construction Methods," *SNAME Transactions,* vol. 107, pp. 291-308, 1999.

[4] M. R. Khedmati, M. Rastani and K. Ghavami, "Ultimate strength and ductility characteristics of intermittently welded stiffened plates," *Journal of Constructional Steel Research,* vol. 65, pp. 599-610, 2009.

[5] M. Razola, "On Structural Design of High-Speed Craft," KTH School of Engineering Sciences, Sweden, 2013.

[6] Ship Structure Committee, SSC-410 Fatigue Of Aluminium Structural Weldments, 2000 ed., Ship Structure Committee, 2000.

[7] Collette, M., 2005. Strength and Reliability of Aluminium Stiffened Panels. Ph.D. University of Newcastle.

[8] F. M. Mazzolani, "Design of Aluminium Structures," University of Naples, Federico.

[9] M. Pedram and M. R. Khedmati, "The effect of welding on the strength of aluminium stiffened plates subject to combined uniaxial compression and lateral pressure," Int. J. Nav. Archit. Ocean Eng, vol. 6, p. 39~59, 2014.

[10] EuroCode 9, Part 2: Structures Susceptible to Fatigue; Design of Aluminum Structures, Brussels: European Committee for Standardization, 1998.

[11] ABS, Guidance notes on Structural Direct Analysis For High-Speed Craft, Feb 2014

[12] DNV, Fatigue Assessment of Ship Structures; Classification Notes No. 30.7, April 2014

[13] Z. D.Perovic, "The Weld Profile Effect On Stress Concentration," in Trends in the Development of Machinery and Associated Technology, Czech Republic, 2011.

[14] N. R. Tom Lassen, Fatigue Life Analyses of Welded Structures, London: ISTE Ltd, 2006.

[15] X. Niue and G Glinka, "The weld profile effect on the SIF in weldments" International Journal of Fracture 1987 (35), pp 3–20

[16] C. Morgenstern, C. Sonsino, A. Hobbacher and F. Sorbo, "Fatigue design of aluminium welded joints by the local stress concept with the fictitious notch radius of rf=1 mm," International Journal of Fatigue, 05 January 2006.

[17] ABS, Guide for Fatigue Assessment Of Offshore Structures, Feb 2014

[18] S. Maddox, "Review of fatigue assessment procedures for welded aluminium structures," International Journal of Fatigue, vol. 25, 2003.

[19] ABS, Rules for building and classing steel vessels, 2014.

[20] ABS, Rules for Building and Classing High Speed Craft, 2016

A MODULAR SEARCH AND SURVEILLANCE SES PLATFORM TO SUPPORT PORT AND HARBOUR SECURITY IN THE FUTURE MEGACITIES OF 2050

S McCartan, EBDIG-IRC, FAH, Coventry University, UK
J Hill, Trident Marine, UK
T H Espeland, ESNA AS, NO
T Dobbins, STResearch, UK

SUMMARY

By 2050 urbanisation will be the most consequential event in the history of mankind, with the rise of numerous megacities primarily located in coastal areas. The plans for any security or disaster operation in a megacity involves a full spectrum of capabilities, spanning both government/military and civil roles in the face of natural or adversarial threats. The security resources of a megacity may also be called upon in more non-security/defence missions to protect or sustain life, such as SAR. This paper discusses the maritime capability required to support the search and surveillance operations within and around the megacity. The overall capability is delivered by an integration of manned, unmanned and autonomous systems, including the use of swarm technology.

Specific domain capabilities include: littoral and harbour areas; riverine and canal operations; underwater operations; urban airspace, with tethered and untethered C4ISTAR sensors; road, cross-country and tunnels. The maritime system proposed in this paper consists of a fleet of modular platform vessels, based on a 25m SES (Surface Effect Ship), operating within a common command infrastructure. Where a range of platforms are deployed from the vessels depending on the situation requirements. The recognised stake-holder informed Marine Design approach is applied to the platform system and sub-systems development, which also draws on the results of horizon scanning activities to identify the technological and cultural developments that are anticipated. These include AI and machine learning, and how they will operate alongside the humans to maintain the security and health of the population.

1. INTRODUCTION: MEGACITIES RISKS AND OPPORTUNITIES

By 2050 urbanisation will be the most consequential event in the history of mankind, with the rise of numerous megacities primarily located in coastal areas. Over half of all people currently live in cities, and the rate of migration is accelerating. By 2030, cities will account for 60% of the world's population and 70% of the worlds GDP. The cities that grow the fastest will be the most challenged. Urban areas are expected to grow by 1.4 billion in the next two decades, with that growth occurring almost entirely in developing world. As resources become constrained, illicit networks could potentially fill the gap left by over-extended and under-capitalized governments. The risk of natural disasters compounded by geography, climate change, unregulated growth and substandard infrastructure will magnify the challenges of humanitarian relief. [1]

Megacities will continue to occupy key strategic terrain, making their stability necessary for global connectedness and order. Regardless of the fragility or resilience of the city, their stable functioning is dependent on systems of finite capacity. When these systems experience demand which surpasses their capacity, the load on the city's systems erode its support mechanisms, increasing their fragility. These systems are then more vulnerable to triggers which can push the city past its tipping point and render it incapable of meeting the needs of its population. Some dynamics of friction are observable in all megacities to varying degrees. population growth and migration, separation and gentrification, environmental vulnerability and resource competition, and hostile actors are all present in some fashion within every megacity.[1]

Cities differ widely on their ability to adapt to volatility and stress. Some cities respond poorly to adversity, making bad situations worse. Others return quickly to a normal state, expending resources to minimizing the impact of the adversity. These characteristics are *fragility* and *resilience* respectively. Many cities learn and grow from adversity, a characteristic that is coming to be known as *antifragility*. There is a strong correlation between highly integrated systems and antifragility. In large urban environments, a highly integrated city like New York exhibits antifragile characteristics when it learns from setbacks and then designs systems that prevent future disruptions from similar events. Loosely integrated cities, on the other hand, show little or no improvement in the aftermath of adverse events. [1]

The ability of megacities to withstand and recover from expected or unexpected stressors is largely based on the city's degree of integration. Some megacities are more integrated than others and, when compared to one another, this degree of integration can be used to categorize megacities into a basic typology, as shown in Figure 1. It ranges from cities that are highly integrated (e.g. New York City or Tokyo) with hierarchical governance and security systems, to cities that are loosely integrated (e.g. Lagos, Nigeria or Dhaka, Bangladesh) with alternatively governed spaces and security systems. Some cities exhibit a combination of the two. Highly integrated systems are characterized by strong formal and informal relation-ships among its component parts. These relationships manifest as highly ordered hierarchical structures with formalized procedures and norms, and open communication among its various parts. Highly integrated systems are inherently stable, show high degrees of resilience (ability to absorb stress) and manage growth in a relatively controlled manner. Loosely integrated cities, on the other hand, lack many of the formal relationships that keep highly integrated cities stable. Weak control and communications systems, and lack of consistent rules for interaction amongst component parts lead to low resilience and unregulated growth. This growth, in turn, contributes more component parts that aren't formally integrated into the system, creating a downward spiral of instability. Loosely integrated cities are largely incapable of dealing with the challenges presented them today and there should be little expectation of their ability to meet the growing challenges of tomorrow. [1]

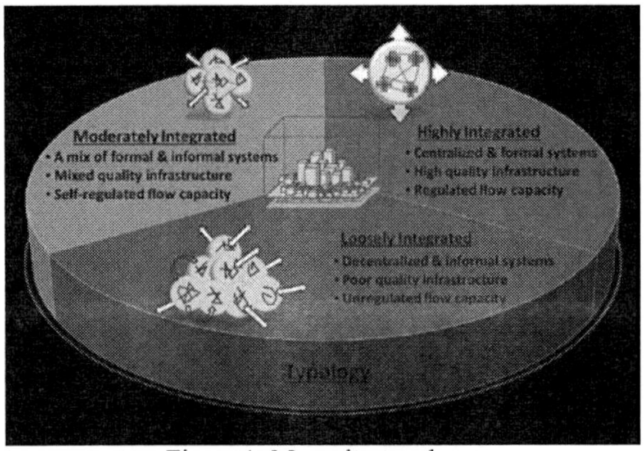

Figure 1: Megacity typology

Some megacities, particularly highly integrated cities, are capable of coping relatively well when instability arises, while others will have their service and security capabilities quickly overwhelmed. The typology defined here provides a basic ordinal categorization to compare cities based on their degree of systemic integration. Viewed through this typology, New York is highly integrated. Bangkok, Rio de Janeiro, and Sao Paulo show moderate degrees of integration, while Lagos, Nigeria and Dhaka, Bangladesh appear far less integrated. New York City (NYC) has become one of the most iconic and important cities in human history, as the home for the United Nations, it is the world's hub for international diplomacy and conflict resolution. It is the epitome of the Highly Integrated city, where its redundant systems, hierarchical structure, robust resources and connectedness make this one of the most resilient cities in the world. Military operations in environments like NYC would likely have a prescribed function, integrated into the cities already robust response framework. The key findings of the analysis of NYC were:

- Keeping NYC functioning is of national and global importance, and, in terms of capacity, the city sets the world standard for integration, antifragility and resilience. The city is densely instrumented and richly resourced.
- As an example of cities eclipsing states or nations in importance, NYC's $1.13 trillion GDP ranks 13th in the world (comparable to Canada and ahead of South Korea and Australia). NYC has its own foreign policy and its own State Department.
- New York has a mixture of different cultures un-matched in the United States. With 600 separate cultural institutions, as well as over 51 million domestic and foreign tourists each year, agencies that provide services and security within the city must be able to understand and serve over 100 different nationalities often living in tightly knit Diaspora communities.
- New Yorkers are globally connected. physically, socially, and technologically. The city enjoys a rich history of assimilation and integration, which contributes to its diverse demography, and is made possible by NYC's transportation infra-structure. NYC houses three international airports and a seaport which is the largest on the eastern seaboard and third largest in the nation. Virtual connectedness permeates the city and typifies the density of handheld communications and web enabled devices in highly integrated cities. This connectedness limits opportunities for tactical or operational surprise.
- The city's heterogeneous population, while connected and integrated into the city, maintains unique sub-cultural identities with strong ties to various countries of origin. These ties are virtually supported through state of the art telecommunications and on-line social networks that enable monetary remittance flows to their families abroad.
- The robust flow systems allow the population in Manhattan to fluctuate between 4 million people during the weekday to 1.6 million on weeknights. These same networks move 6 million people every day throughout the entire metropolitan region.
- The city exhibits anti-fragility. When an attack or natural disaster occurs and a critical flow is disrupted, supporting infrastructure is not simply re-built; it is rebuilt stronger and able to withstand more. [1]

1.1 MEGACITY INFRASTRUCTURE CHALLENGES

A report by MRC McLean and Globalscan [2] studied infrastructure challenges and governance trends in the world's largest megacities in 2006. The research focuses on five critical infrastructure sectors: transportation; water; electricity; healthcare; safety & security. It also looked at how city management is evolving to address these challenges, based on the views of a range of municipal stakeholders. Their key findings on safety and security were:

- Safety and security comes second only to transport as a factor contributing to city competitiveness
- Organized crime emerges as the biggest problem, followed some distance behind by terrorism
- Cities are attempting to shift to a more proactive approach to security
- Although open communities are the favoured as the route forward for most, gated communities will also be emphasized by many, especially in emerging cities
- Public surveillance is deemed more important than protecting privacy

The complete survey population ranked public safety and security as the second most important infrastructure area in determining competitiveness. It is also the sixth-biggest social challenge, while crime/corruption comes in tenth. Threats against the city: The more detailed survey answers by security stakeholders elucidates their more informed concerns. Organized crime, including by armed gangs, is their biggest challenge. Next, a surprising distance behind, comes terrorism a particular concern in emerging and mature cities. Although distinct threats, the two overlap in important ways. Terrorists use organized violence to weaken the state: criminal gangs do so to gain wealth. In practice, the main problem facing security professionals is organized groups that challenge the rule of law, and that increasingly cooperate with each other whatever their goals. The next most pressing safety concern, potential natural disasters, shows the importance of context in risk perception. [2]

In terms of how best the city could address security issues, the most popular response is additional officers and law enforcement capacity while the third is better preparedness and planning. The common message was more capacity and better used. Over recent decades numerous police departments have shifted from a reactive approach to take the fight to the enemy. The survey results show this occurring worldwide. The success of so-called "risk-based policing" has been dramatic. New York City's ongoing, widely copied COMPSTAT program, which analyzes police data to indicate crime hotspots, allowing frequent tactical resource reallocation, was important in that city's drop in crime figures in the late 1990s. [2]

A strategy based on predicting crime has its controversial elements. Survey respondents thought that the emphasis on the need for public surveillance would far outstrip privacy concerns, a conviction strongest in transitional and mature cities, where risk-based policing is most advanced. Technology is essential for risk-based policing, and a vital tool for general surveillance in cities wishing to increase police capacity. Crime mapping is just one example. Another ubiquitous and sometimes very useful one is CCTV. The security stakeholders see technology's importance, expecting the emphasis on its purchase to be greater than investment in human resources. Technology is a crucial tool, not a complete solution. [2]

1.2 URBAN STAKEHOLDERS

In terms of urban stakeholders and the implications of urbanization, Governments, businesses, citizens and civil society will need to work in synergy to address urban challenges. As urban demands and complexity of infrastructure and services grow, the resources required to tackle them are increasingly spread across many different stakeholders. In this scenario, partnerships and "co-production" are developing more relevance. Global citizens are seeking enhanced interaction and multi-layered experiences, with technology the key enabler of cultural exchange and engagement. Citizens are emphasizing access over ownership, have a keen interest in sustainability issues, expecting goods and services to be produced and delivered responsibly. Government agencies can capitalize on the increased awareness and accessibility of urban residents to elicit a higher level of engagement and community participation, to facilitate co-creation and to reduce the community risks in urban development initiatives. [3]

The complex operation of cities involves a wide range of domains such as mobility, infrastructure, social services, the environment, knowledge and skills, and culture. It also involves a large number of stakeholders in the planning and administrative process, including national, regional and local governments, and a number of beneficiaries and partners, such as citizens, the private sector and NGOs. City planning and administration have changed significantly in the last decade with the rise of technology; disruptive business models and technological solutions can enable the sharing of data across domains and help cities build situational awareness. How urban infrastructure and services are being managed has attracted increased private-sector and entrepreneurial interest. Cities are constrained by both budgets and capacity to offer services over digital channels. The private sector is enabling cities to adapt faster, using its innovation and agility through several models including Digital "integrated city" services.

An integrated view of all urban services is required to effectively manage private provisioning of services. The City of Madrid implemented an intelligent technology platform called as MiNT (Madrid Intelligence) that allows for management of service providers and communication with citizens. A key feature of the solution is a mechanism where citizens can provide immediate feedback about a service or event in the city with their location details. The platform integrates data from citizens with data captured through sensors, cameras and IoT (Internet of Things) devices to generate a comprehensive view of city services. The platform is built on a database of more than 5 million city assets and provides a real-time city view, enabling better responses to service requirements. The platform provides a mechanism to measure and analyse more than 300 performance indicators across urban domains and incentivizes service providers through better compensation for higher quality of delivery. This contributes to improved quality of life in the city as private agencies responsible for management make their best effort to improve service benchmarks for better compensation. Datasets created on account of city functions hold immense value and can be used to constantly analyse and improve city operations across all domains. The case study highlights the benefits of progressive technology in enhancing urban services, and that transferability is high for cities that have a mature urban services system and data available on core infrastructure and services. [3]

1.3 JOINT URBAN OPERATIONS

The U.S. Department of Defense's (DoD) Roadmap for Improving Capabilities for Joint Urban Operations (JUO) is predicated upon continuous ISR efforts with fire delivery and ground manoeuvre, whether by special operations forces, or by conventional ground forces, or their combinations. The concepts include precision strike, nodal capture, nodal capture and expansion, soft-point capture and expansion, segment and isolate, and nodal isolation. Where precision strike involves precise indirect and direct attacks to achieve desired effects against adversary capabilities from standoff distances, without occupying ground. It requires precisely knowing locations of nodes of adversary forces and how they interact. The joint force commander minimizes ground force presence by using remote fires and Special Forces direct action as his primary defeat mechanisms. Here, as with all concepts, operations require a joint, interagency, intergovernmental, and multinational (JIIM) operational approach. Moreover, a light footprint is required with heavy reliance on leveraging existing systems in the city. Reliance on current technology within the megacity presents a challenge, since dense urban infrastructures make it difficult for US forces to fully employ long-range sensors and munitions. Moreover, civilian populations are an ever-present reminder of the need to avoid collateral damage. Overall, a reliance on technology and "constant ISR" permeates schemes for operations within megacity. No ISR platform is perfect, but megacities demand improved platforms. The megacity will require rapid sharing of information over dispersed formations and platforms capable of manoeuvring within and providing near-real-time intelligence in the megacity. [4]

In terms of conceptual thinking and layered intelligence for urban military contingencies, Williams and Selle [5] conceptualisation of any city as a living organism is useful in terms of assessing the relationship between the city and the environment. Specifically, in the context of delineating intelligence needs for military contingencies in megacities and sub-megacities. This is shown in Table 1, which highlights the parallels between humans and the city as as an organism.

Human Entity	City as an Organism
Skeleton	Physical infrastructure
Core and Periphery	Centre and peri-urban
Neural networks	Social networks
Metabolism	Flows
Blood flow	Pulse or rhythm
Brain functions	Governance
Survival instincts: fight-or-flight mechanisms	Coping mechanisms: informal or illegal economies
Immune system: antibodies	Resistance to external intervention

Table 1: Comparison of a Human to a City as an Organism

Treating the city as an organism helps to bound the intelligence process and identify key elements, such as social networks, flows, governance, the rhythm of cities, and the likely response to external intervention, even when it might be benign. The idea of the city as an organism has also become important in understanding some of the results of the increased size of cities. There are systematic scaling laws which explicitly show that cities are more than the linear sum of their individual components. For example, economic productivity increases systematically on a per capita basis by 15% with every doubling of a city's population, regardless of a city's initial size.

Cities are a layered and interacting series of complex adaptive systems involving actions, interactions, and transactions. This also requires multiple streams of intelligence: information on network and power structures could be obtained and updated from the intelligence community; infrastructure mapping, breakdowns from satellite data, and considerable social, economic, and political intelligence could be obtained from non-governmental organizations. All of this would lead to an enhanced understanding of urban dynamics, including patterns of flows and networks of relationships. Operating effectively in megacities requires an understanding of these flows and networks. This focus and approach can be extended to facilitate the understanding of governance patterns and governance processes in urban areas. Indeed, intelligence preparation of and on the urban battlefield is based on 10 different layers of the city, recognizing that these are not independent from one another, but are constantly interacting. There are clearly multiple sets of vertical connectors between the various layers, both direct and indirect. Some of these connectors will have readily predictable consequences, while others will be much harder to assess. Nevertheless, by treating these as a series of layers, it should be possible to elucidate the need for a comprehensive approach to intelligence in megacities.

In thinking about intelligence for military contingencies in urban areas, it is important to consider it as a dynamic interactive process. Intelligence preparation of the battlefield will be vital prior to action; intelligence preparation on the battlefield will be a continued and even more stringent requirement throughout the duration of the contingency. The U.S. military will have to be able to effectively piece together a comprehensive and actionable intelligence picture, and under enormously challenging circumstances. This would require, an ability to persistently monitor, collect, and interpret, in near real-time, the millions of bits of data associated with cellphone communications, social media postings, financial transactions, and the operational movements of these actors. The challenges associated with doing so will likely require the U.S. military to develop the ability to manage and interpret in a timely fashion the unending stream of data. The levels or slices of the intelligence challenge are identified in Table 2.

Level 1	Subterranean
Level 2	Topography—roads, chokepoints, etc.
Level 3	The Cityscape—the buildings
Level 4	The Service Infrastructure
Level 5	The People—human terrain
Level 6	The Networks—social or criminal capital
Level 7	The Flows—people and things
Level 8	Forms and Spaces of Governance
Level 9	The Rhythm of the City
Level 10	The Cyber and Electronic Layer

Table 2: Levels of intelligence challenge

Gentile et al [6] conducted a historical analysis of the ways in which militaries have deployed light and mechanized infantry during close urban combat. The objective was to identify the comparative advantages and costs of this warfighting approach and to distill the lessons that might be gleaned from these experiences. These lessons can also inform how the Army might confront similar foes in complex, urban environments in the future. This analysis focuses on five case study battles: the U.S. Army in Mogadishu in 1993, the Russian Army in Grozny in 1994 and in 1999, the U.S. Army in Baghdad in 2003, the U.S. Marine Corps and U.S. Army in Fallujah in 2004, and the U.S. Army in the Sadr City suburb of Baghdad in 2008.

The analysis identified that the Russians failed to appreciate was just how much the urban environment favours the defender. Figure 2 depicts many of the features of the urban landscape that enemy forces and sympathetic locals who are familiar with the terrain can exploit, from attack positions above and inside multistory buildings to maneuver routes in subterranean underground sewer systems below.

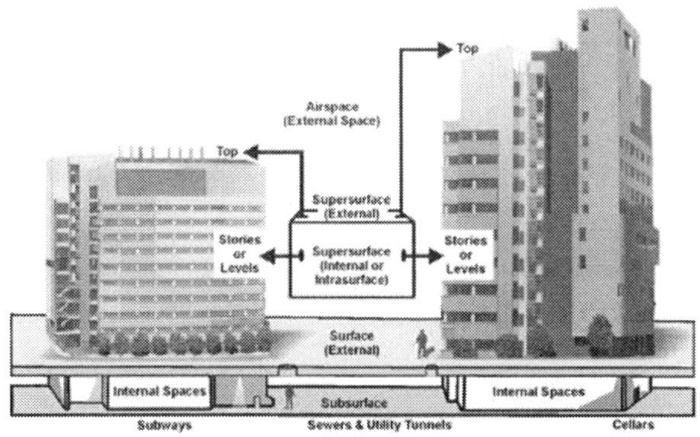

Figure 2: Features of the urban landscape

These were locations that Russian vehicles could not engage because their tank guns could not elevate or depress sufficiently. Additionally, as with the combat vehicles of most armies, Russian vehicles were designed for head-to-head combat with other armoured vehicles, and what seemed to be reasonable design trade-offs about protection from top, side, and rear attacks were made to accommodate conventional challenges. Chechen fighters took advantage of these vulnerabilities in 1995. During the operations in Grozny in 1995, poorly trained Russian soldiers had attempted to force a *coup de main* on the city, relying heavily on armoured vehicles, including tanks. Yet, conventional mechanized forces were no match for Chechen small guerrilla units that could fire off a few rounds and quickly drive off to avoid counterfire. The city's narrow, winding streets enabled snipers to target and immobilize the front and rear vehicles of Russian tank columns, rendering the middle vehicles easy prey for RPGs. Additionally, the main guns of the tanks could not elevate or decline enough to engage and counterattack these rebel positions.

As a result, the Russian Army struggled and ultimately failed to control the outer layers of Grozny on the way to the city centre, enabling resistance fighters to manoeuvre and resupply as needed. The Russians did not conduct a rigorous IPB, which would have highlighted such critical nuances about the urban terrain. It also would have revealed the proliferation of RPGs and anti-aircraft weapons and talented snipers, and the enemy's proclivity and acumen for ambush warfare. Additionally, the lack of an adequate IPB produced infantry soldiers unfamiliar with the language and indifferent to local culture.

2. ANALYSING THREATS, POLICIES AND SOLUTIONS IN MEGACITY PORT AND HARBOUR SECURITY

Port and harbour security became an issue of growing concern in the wake of the September 2001 (9/11) attacks on the US, the attacks on the naval destroyers USS The Sullivans and USS Cole while tied up in the Port of Aden in Yemen in 2000, and the French oil tanker Limberg in 2002. Prior to these attacks, the main focus of port security was often directed at landside threats such as the theft of cargo from containers, warehouses, or ships berthed in port. Securing perimeter fencing to prevent thieves from accessing the port area including warehouses was often the main concern, particularly in ports adjacent to urban areas. Similarly, there is potential for undocumented migrants to enter a port and attempt to stow away on ships, in containers or in trucks parked in the port area.[7]

In the aftermath of 9/11 international measures were introduced by the International Maritime Organization (IMO) to provide greater protection from potential terrorist attacks on ports and on the ships using them. These measures included the International Ship and Port Facility (ISPS) code, a set of measures introduced in 2002 under chapter XI-2 of the International Convention for the Safety of Life at Sea. This establishes mandatory requirements on both ships and ports to enhance maritime security. A code of practice on security in ports was introduced by the IMO and the International Labour Organization, in 2003.[8] This non-legally binding instrument was designed to provide guidance to all those responsible for addressing the issue of security in ports. The aim of port security measures is to maintain an acceptable level of risk at all security levels. Where security measures should be devised to reduce risks and should in the main revolve around procedures to establish and control access to restricted areas and other vulnerable or sensitive key points, locations, functions or operations in the port. The aim of security measures include preventing access to the port by persons without a legitimate reason to be there and prevent those persons with legitimate reasons to be in the port from gaining illegal access to ships or other restricted port areas for the purpose of committing unlawful acts. The following security levels give examples of security measures to be adopted:

- **Security level 1** measures may include random personnel, baggage, material and stores and vehicle screening, and implementation of access and movement control.
- **Security level 2** measures may include increased frequency of screening, more robust monitoring of the port, and more stringent access and movement control measures.
- **Security level 3** measures may include 100 per cent screening, increased identification checks, temporary cessation of certain port activities and/or imposing vessel traffic control measures, restricting access to certain areas, deployment of security personnel to key infrastructure, etc.

The port security assessment includes the identification of threats to assets and infrastructure, in order to establish and prioritize security measures. An example of a method and risk-based tool to assist in preparing a port security assessment is the "Threat and Risk Analysis Matrix" (TRAM). Its purpose is to identify threats with a view to initiating and recommending countermeasures to deter, detect and reduce the consequence of any potential incident should it occur. Such analysis is a critical aid to allocating resources, forward planning, contingency planning and budgeting. The port security plan (PSP) is based on the PSA.

3. SYSTEM CONCEPT DEVELOPMENT

The system concept uses a fleet of SES operating C4IDR based at Ellis Island, which is uniquely positioned to provide security and SAR capability to the ports and harbours of: Manhattan; Brooklyn; New Jersey; Staten Island. Single bridge access and distance from the mainland strategically enhances the security of the operational base. The proposed system concept development is predicated on a public-private partnership with a Systems Integrator in the Homeland Security sector capable of providing Security Solutions for a wide spectrum of requirements covering border, critical infrastructure, coastal, anti-insurgency, etc. It is based on the design of innovative and cost effective solutions, coupled with state-of-the-art sensors/platforms for harbour and port environments. Where a multidisciplinary approach to system design and field integration are essential. The SES was selected as a high speed platform 40knots+ which could function well in both SAR and security roles, it is discussed in section 4.

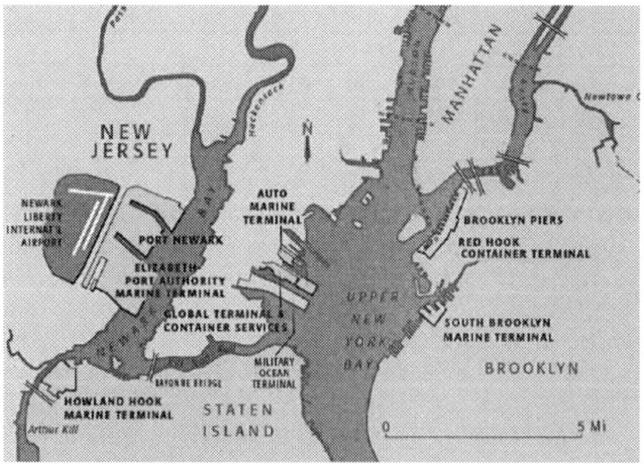

Figure 3: Ports of New York and New Jersey

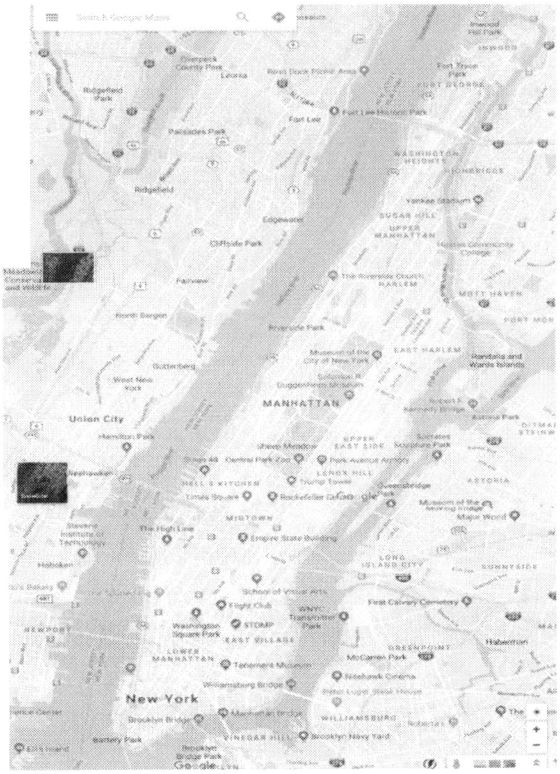

Figure 4 : Manhattan Island

3.1 SCENARIO

Ports remain the primary gateway for international trade and the security of these ports cannot be understated. As economic centers of gravity, ports are a high value target for terrorists and armed forces. The maritime environment has its own challenges, in which traditional surveillance methods may not be effective. Ports require a multi-layered security approach to achieve an effective security shield with a common operational picture. This involves selecting and integrating individual best-of-class COTS-based technologies. Requiring an open-architecture approach to facilitate future system growth through multi sensor fusion in the seamless integration of C4i to achieve total domain awareness. The critical asset in this scenario which will receive significant innovation by 2050 is the patrol boat, which is the focus of this paper. This vessel will be based on a modular platform to address both security and SAR roles.

Threat detection on the waterfront itself is different from those in other critical sectors. Under water detection systems, surveillance systems and identification system would require to be linked to command and control system to enable appropriate responses. Port Security requires a solution, which has reliable day and night, all-weather surveillance to detect ships and small crafts at a distance of several nautical miles while allowing smooth integration with existing sensors such as radars, AIS, visible cameras etc. Thus requiring a 360-degree solutions for over the surface, perimeter, underwater and aerial surveillance. When designing a protection system the task should be approached from the four principles: Detect; Determine; Delay; Defend. These principles facilitate the application of appropriate technology in the protection of an area.

Detect – is the process of identifying specific activity that is to be alerted to the security team. Detection should be made at the earliest opportunity. Detection technologies include radar, fibre optic sensors and infrared beams. Water interceptor sensor arrays greatly enhance detection capability with a patrolling fleet of vessels

Determine – is the ability to verify the detected event so that erroneous activations are eliminated and valuable intelligence is gathered relating to real events. Determination technologies include long range CCTV and thermal imaging. Water interceptor sensor arrays greatly enhance verification capability with a patrolling fleet of vessels able to rapidly interrogate a potential threat at close quarters.

Delay – is important to allow a security response to be initiated. This may be a physical barrier or an open area that will take time to cross. Delay technologies include fencing, gates and barriers.

Defend – is the deployment of technology to deter or repel a security breach. Defence technology includes power fence and electronic jammers. On the water and at coastal areas water interceptors can rapidly deploy troops and vehicles to win the fire fight and secure the area or vessel that poses a threat.

An Integrated Port Operations Systems (POS) is a single umbrella ensuring the secure and efficient flow of goods through the port while managing port facilities to maximum efficiency. This paper will focus on the maritime security aspects of this. The integration platform (POS) is an holistic management system covering security, port community and facility management. It is designed to integrate multiple systems into a Common Operating Picture (COP). The POS resides on top of all the other systems with a single interface for monitoring, controlling and reporting on all the sub-systems. Centrally monitoring multiple systems: Radar, Sonar, VTS/AIS, Access Control, Video Management, SCADA, Utilities Management, Fire & Security alarms, Life Safety, Intrusion Detection, Intercom, Mass Notification, Network Activity & many others.

Port security management sensor technologies for the sea and land perimeters are shown in Figure 5. A typical port will require different zones, as shown in Figure 6, with integrated surveillance fence technology, such as: radar; sonar; microwave; thermal camera. Resulting in a complete surveillance picture in command & control surveillance & tracking of: sea-surface targets (small boats to large ships); sub-surface targets (divers, submersibles); intruder detection & identification.

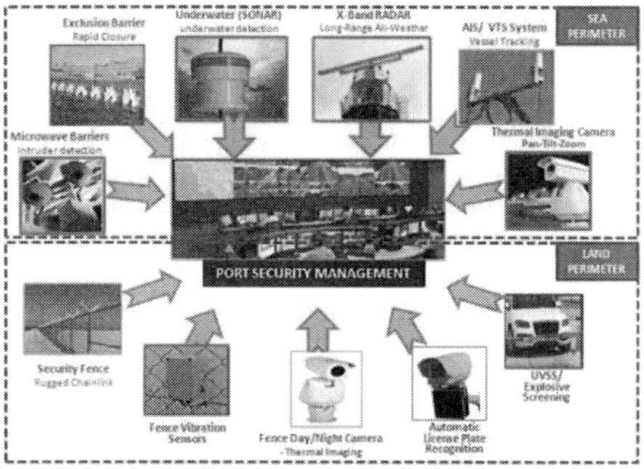

Figure 5: Sensor technologies for the sea and land [9]

Figure 6: Typical port showing different zones[9]

The land perimeter is shown in Figure 7, which consists of a high security complete fence with high-integrity crash-proof security gates, located at the major entry points into the port facility. Perimeter intruder detection is provided by an array of thermal imaging day/night cameras, fence vibration sensors and tripwires, which use buried volumetric sensors and microwave solutions. There are explosives detection measures integrated with security screening procedures, in place at the security gates based on: under vehicle screening systems; automatic licencse plate detection and recording; hand-held explosive detectors; under-vehicle explosive inspection system.

Figure 7: Land side perimeter with security gate [9]

X or S-Band long-range radar is an all-environment capable technology for vessel surveillance and tracking, shown in Figure 8. Where multiple surveillance radars are used for all-round surveillance, providing multiple target tracking with a range up to 30km. It is supplemented by a long range thermal camera for target ID, co-located on tower with radar, which has a surveillance capability up to 25km for man-in-sea.

Figure 8: Location of radar on tower [9]

The underwater sonar barrier, deployed at entrance to port for detection & tracking of all difficult to detect underwater targets across all sea-states, is shown in Figure 9. This fully automated underwater security solution with a range up to 1500m (>7km2 coverage) per sonar head, active and passive detection and Classification Detects all potential sub-sea targets: divers; submersibles; submarines. The Automatic Identification System (AIS) base stations are mounted above the water at the same location. They transmit/ receive for ship's identification/ position & tracking. The integrated C2 system automatically identifies and locates vessels by electronically exchanging data with other nearby ships & AIS Base stations. Supplements marine radar for collision avoidance.

Figure 9: Location of underwater sonar barrier [9]

Microwave barriers are used for the jetties, providing exact location of intruder, with slew-to-cue to camera for positive ID. Where transmitter/receiver pairs using microwave beam in X-Band, are located every 200m. The harbour protection barrier, shown in Figure 10, is a Vessel Exclusion Barrier (EVEM), which can rapidly be put into place at the entry to a port. It creates a significant barrier to intrusions and/or entry by unauthorized vessels, with high-strength capture net designed for vessel incapacitation.

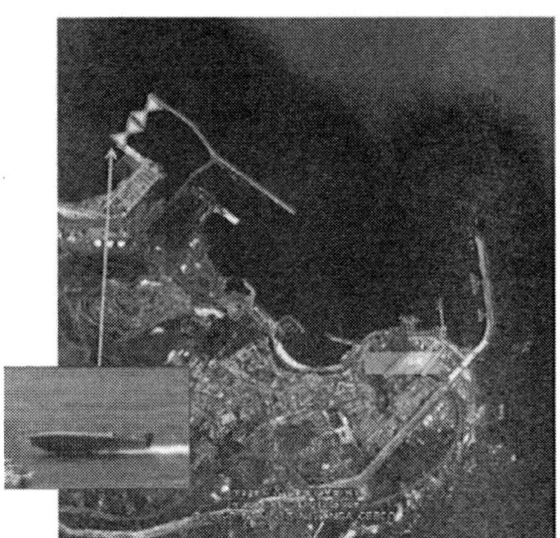

Figure 10: Harbour protection barrier [9]

3.2 COMMAND-AND-CONTROL SYSEMS

The development of a command, control, communications, computers, intelligence, surveillance, and reconnaissance (C4ISR) architecture based on the Internet Protocol, a service-oriented architecture (SOA), and the developing capabilities of the Global Information Grid (GIG), promise to provide important elements of the flexible C4ISR needed for flexibly constituted naval strike groups. This section focuses on the command-and-control (C2) portion of C4ISR and on the relationship of Navy research and development programs to the vision of network-centric operations (NCO).[10]
The status of naval C2 systems development including efforts related to establishing a common operational picture (COP) is reviewed first. Research efforts within the U.S. Navy and the Department of Defence (DOD) addressing C2 systems employing SOAs and related advanced concepts are then reviewed.

A key technology for 2050 will be the Distributed Common Ground Station (DCGS), which is envisioned to be a family of systems that provides multi-ISR processing and exploitation to the Joint Task Force (JTF) and echelons below. The DCGS Integration Backbone (DIB) and NCES share many common, desired characteristics. DIB and NCES efforts would collaborate to provide common products, thereby helping to establish closer coupling between C2 and intelligence, surveillance, and reconnaissance (ISR) processing systems.

3.2.1 COMMON OPERATIONAL PICTURE

Tracking data from numerous sources, including weapons systems, organic and nonorganic sensors, and intelligence sources are inputs to the C4i system to generate the COP as an output. The COP consists of: air picture; maritime picture; undersea picture; ground picture. An accurate COP is essential to NCO, as it facilitates the self-synchronization of NCO, decreasing the need for communications to establish a common understanding of a situation and thereby increasing the speed of command. In a COP all participants have the same display picture; rather, it means that all participants have access to common sources of data, which could be displayed in different ways depending on the needs and equipment of the particular user. Access to data is the key here. From a network-centric perspective, users should have access to data as soon as they are in some comprehensible form, even though further processing of the data might be intended. This is because different users will have different needs for the data, and the additional processing might remove information content according to the perspectives of some users. [10]

3.2.2 COMMAND AND CONTROL WITH SERVICE-ORIENTED ARCHITECTURES

Service-oriented architectures (SOAs) is a relatively new concept. SOA is a prospectively powerful technical infrastructure for composing forces "on the fly" to respond rapidly to new threats and to streamline the force buildup required to conduct military operations. Building this technical infrastructure for network-centric operations is an exceptionally large undertaking, which as a next generation technology would be fully developed by 2050. The U.S. Navy has several ongoing research programs related to SOA for C2, as follows:

- Composable FORCEnet—SPAWAR,
- FORCEnet Engagement Packages—SPAWAR,
- Extensible Tactical C4I Framework—ONR,
- Enterprise Services Bus (ESB)—SPAWAR, and Joint Coordinated Real-Time Engagement (JCRE)—ONR.

Composable FORCEnet is the ability to select on the fly from a vast network specific information resources that are best suited to solving a particular problem or providing specific information, shown in Figure 11. [10]

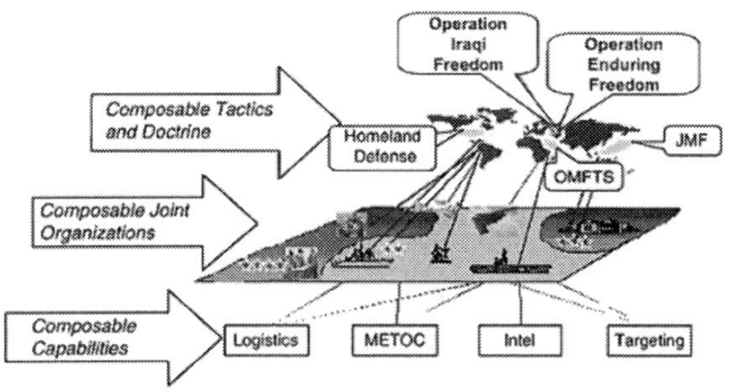

Figure 11: A representation of composable FORCEnet [10]

The implementation of a composable FORCEnet must provide several overarching capabilities:

- The ability to identify participants in the battlespace and to determine their organic capabilities and needs.
- The ability to assemble the participants within a coalition working toward a common mission objective.
- The ability to exchange information within the coalition. ONR is developing the XTCF to provide this capability.
- The ability to establish new concepts of operation (new ways of doing business) that are consistent with the capabilities and constraints of the participants.
- The ability to introduce new capabilities to the coalition, either by adding functionality to existing coalition members or by adding new participants to the coalition

SPAWAR has conducted some early analysis on the mission value of a rapidly composable force for ad hoc warfare. The preliminary results indicate increased combat reach in selected scenarios. This could be applied as a Transfer of Innovation to the security sector.

A FORCEnet Engagement Package (FnEP) is a portfolio of the capabilities that participants bring to a fight. A portfolio is linked to specific mission areas that the participant(s) can support. Participants, here, can range from single entities to multiple entities constituting a larger organization. Figure 12 illustrates a specific FnEP for a notional collection of participants. The rows represent distinct participants; the columns represent capabilities and the specific subsystems that provide them. It shows an arrangement of four participants, all providing sensors, two providing weapons, three providing C2 capabilities to different degrees, and two providing targeteers and/or weapon controllers. The wiring shows the information flows between the participants and the subsystems. The FnEP is defined by both the platform types (including their organic capabilities) and the wiring between them. A different wiring arrangement and/or a different collection of participants would define a different FnEP. [10]

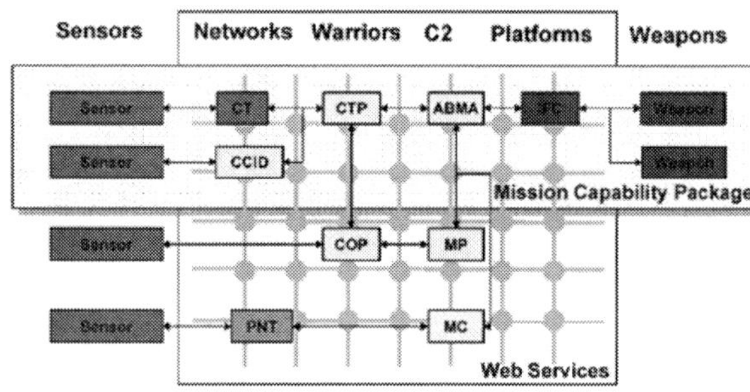

Figure 12: A FORCEnet Engagement Package [10]

The FORCEnet Engagement Package concept is in its formative stages. It will be a fully implemented process by 2050. The concept is dependent on additional capabilities that are themselves only beginning to emerge or mature, including the following:

- The ability to discover the available participants on the fly through a registry.
- The ability to compose an Engagement Package.
- The ability to orchestrate the execution of the Engagement Package.

Enterprise Services Bus

The Enterprise Services Bus is the product of SPAWAR's Distributed Services Commercial Area Announcement. The ESB is a design proposal, for integrating an SOA with field-configurable work-flow management. The ESB provides a set of core, enterprise, and mission-specific services:

- Authentication and authorization services, which isolate identity-management solutions from the bus
- A registry service for registering distributed services
- A publish-and-subscribe messaging/alert service for information flow
- A work-flow orchestration service based on a commercially available BPEL server from Collaxa, Inc.
- Commercial and open-source portal products

- A set of legacy and mission applications comprising mission planning, medical, strike packages, hazardous plume analysis, logistics, and weather information

- A set of authoritative data sources for medical information and blue forces intelligence.

One of the transformational capabilities provided by the ESB is the work-flow orchestration service. This service, implemented using BPEL for Web Services (BPELWS) is based on web-enabled work-flow management standards originally developed by IBM (Web Services Flow Language, WSFL) and Microsoft (XLANG). Operational users can compose their work-flow rules to describe the activities that must be performed to execute a mission thread, the services that they require in order to carry out those activities, and the preconditions, postconditions, and triggers for those activities. Services and triggers can link to other mission threads, thereby making it possible to compose system-level (e.g., platform-level) work flows from component-level work flows, and family-of-system-level (e.g., force-level) work flows from system-level work flows. Operational users can compose these rules at any time: during the workup of a force, during the deployment of a force, or in real time. The binding of the abstract services defined in the work-flow rules to specific services is done at runtime using service discovery and service access methods. [10]

Figure 13 illustrates a BPEL work flow for a representative target-develop- ment process that culminates in a decision about whether or not to engage a prospective target. The objective of this thread is to assemble a comprehensive strike plan, including persistent ISR support. Work-flow rules determine which activities (circles) need to be executed and when, including branching points to parallel activities and convergence (collection) points to serial activities. The specific services that execute those activities—for example, confirming target identity—are discovered at runtime, and are shown as shaded. There does not appear to be any commonality between XTCF and the ESB other than their use of commercial standards for Web services. [10]

Joint Coordinated Real-Time Engagement

The objective of the JCRE, an Advanced Concept Technology Demonstration (ACTD) (Program Element 0603235N), is to demonstrate concepts for force synchronization, both for permission execution targeting and for targets of opportunity discovered on the fly:

- Define mission needs and time lines,
- Define resources required,
- Request resources, and
- Approve or disapprove resource requests.

JCRE provides the operational concepts and software to enable joint real-time operations and engagement across multicombatant command theaters and echelons. It will permit Joint Force Commanders to synchronize and employ military efforts rapidly and effectively to conduct globally time-constrained operations. Figure 14 illustrates the JCRE implementation concept. JCRE provides the following:

- Global Situational Awareness Services.
- Global Resource Management Services
- Global Synchronization Services.

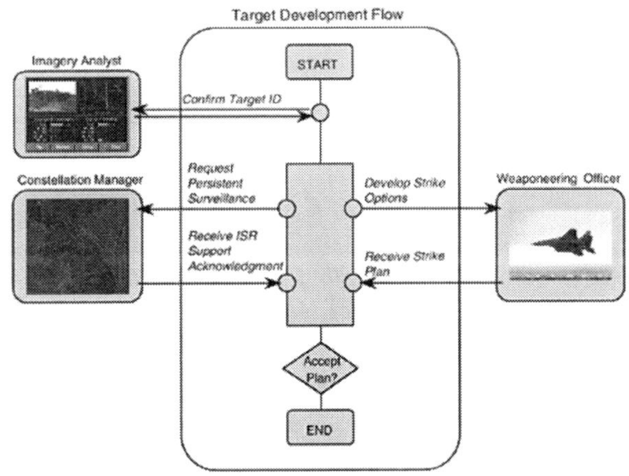

Figure 13: A target-development work flow expressed using BPEL. [10]

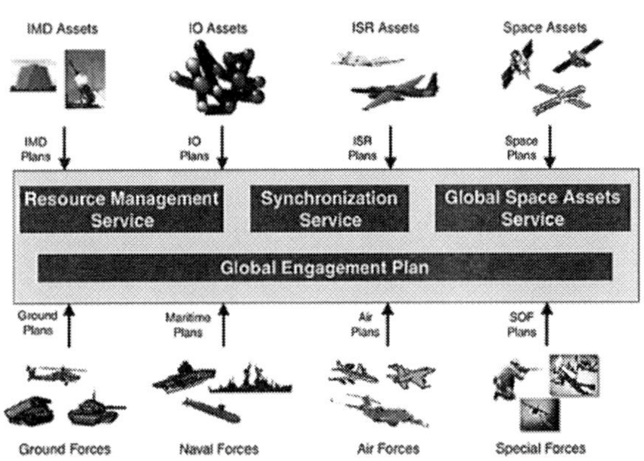

Figure 14: Joint Coordinated Real-Time Engagement [10]

4.4.2 Beyond Service-Oriented Architectures: Model-Driven Architectures and Algorithms

Service-oriented architectures address a fundamental C2 requirement: that of allowing users to configure systems rapidly, employing a collection of pre-existing services. However, what if the available services do not satisfy the needs of the user but need modification? The Object Modeling Group (OMG) has been developing a technology referred to as Model-Driven Architecture (MDA), that addresses this problem: it permits the rapid modification of applications software deployed across a variety of computing platforms. DARPA is taking this approach a step farther, allowing C2 of new capabilities (sensors and weapons) employing new concepts of operations without any modifications to the source code.

The MDA approach shows significant potential, having been successfully employed in several large-scale commercial and military systems and is being used by the Joint SIAP Systems Engineering Office in the SIAP program. However, the MDA technology base is still evolving, and standards have not matured to the point that different vendors' tools are fully interoperable.[10]

The DARPA/IXO Joint Air/Ground Operations: Unified, Adaptive Replanning (JAGUAR) program is also using a model-based approach to achieve architectural agility. The objective of JAGUAR is to transform the pace of operations at air operations centres to "the speed of thought." JAGUAR is embedding knowledge-based plan-development capabilities within a stochastic, dynamic control framework, to create a system that is self-aware, adaptable, and agile, and that scales to large problems and intricate domains. [10]

The idea is to capture in models everything known about entities in a battlespace: how they move, how they interact with other entities, how they are vulnerable, how they are secure, and so on. Once calibrated to respond as its real-life counterpart, a model of an entity can be joined with models of other entities to carry out cooperative tasks, for example, to defeat threats.

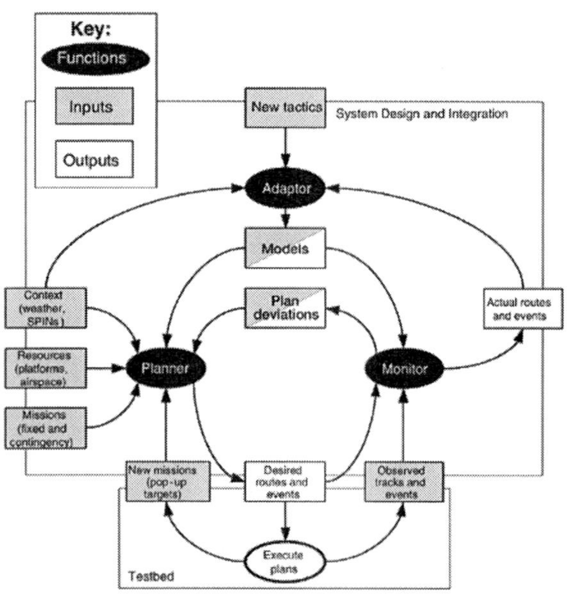

Figure 15: DARPA's JAGUAR program [10]

The most striking aspect of a model-based system is its ability to discover, on its own, new and novel ways of accomplishing tasks. One does not script in a tactic or a rule to accomplish a task in a model-driven system. Rather, one specifies the desired end-state of the task, the constraints in getting to that end-state, the resources (including time) available, the environment (including neutral and deliberately hostile entities), and models for how they all interact. JAGUAR's dynamic control framework, shown in Figure 15, discovers ways to obtain the task end-state, sometimes "discovering" approaches that are used by accomplished human planners and sometimes discovering approaches that have never been practiced. [10]

4. HOST PLATFORM DEVELOPMENT

The maritime system proposed in this paper consists of a fleet of modular platform vessels, based on a 28m SES (Surface Effect Ship), operating within a common command infrastructure. The hull of this ESNA Tern SES CTV platform consists of two narrow side hulls. Between the hulls in the bow there are flexible fibre reinforced rubber fingers. At the stern there is a fibre reinforced inflated rubber bag between the hulls. Centrifugal fans blow air into this enclosed space, providing an air cushion that lifts up to 80% of the vessel weight. The remaining 20% is supported by side hull buoyancy. This allows for the following advantages:

- High vessel speed because the frictional resistance is significantly decreased.
- Decreased wave induced motions because there is less contact with the waves
- The air cushion acts as a large shock absorber, improving the seakeeping and reducing seasickness.

The Tern is equipped with a state-of-the-art motion damping system which further reduces motions both at high and low vessel speeds. As the lift fans supply air to the air cushion, the control system actively controls vent valves that either ventilate the air cushion to a low pressure or closes the valves so that the air cushion pressure is increased.

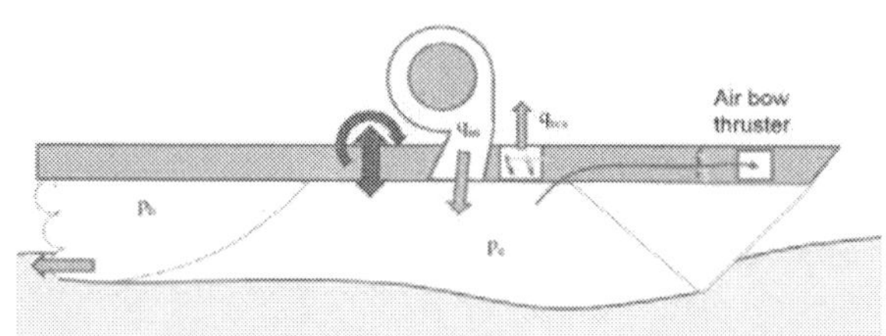

Figure 16: Schematic of air cushion system

The vessel can be lifted up and down approximately 2m, it is thus effectively able to remove up to 2m of vessel motion in waves. This is especially effective in long waves/swell, where traditional vessel types will float and follow the wave surface. This principle is shown in figure 16. While both the CTV (a catamaran with same length as the ESNA Tern) and SOV (a significantly larger single hull vessel) will follow the waves at long wave length, the air cushion control system reduces the vertical motion of the ESNA Tern. At high speed the motion control system smoothens the motions, thus improving the comfort of the passengers and crew, as illustrated in figure 17.

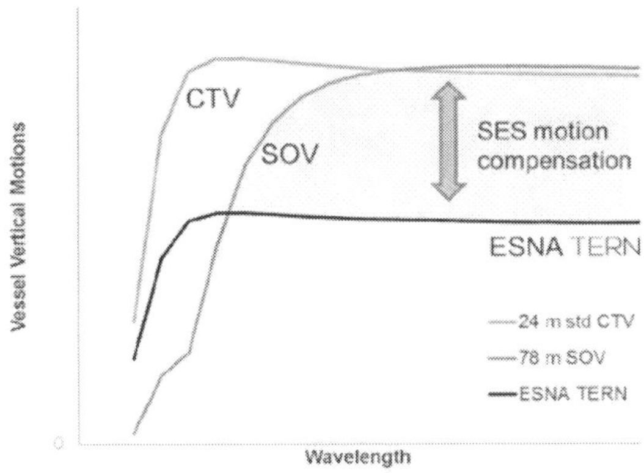

Figure 17: Comparison of vertical motion of SES and other traditional vessel types

MAIN DATA
Structure Aluminium alloy or composites
Length over all: 28.1 m
Beam over all: 10.6 m
Draught: 0.9 m on cushion
3.0 m off cushion

DEADWEIGHT/TONNAGE
Max deadweight 20 t
Net tonnage 41

CAPACITIES
Fuel oil: 10 m3
Max deck load 1.0 t/m2

PROPULSION SYSTEM
Waterjets driven by single shaft/diesel engines with vertical offset gearbox 2 x 1080 kW
Optional: 2 x 1440 kW

SPEED PERFORMANCE

	2 x 1080 kW	2 x 1440kW
Speed (max)	33	40-44 knots
Speed (SS3/1 m Hs)	28	35-39 knots
Speed (SS4/2.0 m Hs)	23	30+ knots
Speed (2.5 m Hs)	21	25 knots

FUEL CONSUMPTION
620 l/hour at 2 x 1080 kW main engine power, one lift fan at 250 kW and 20 kW genset power.

SES SYSTEMS
Diesel engine powered centrifugal lift fans: 2 x 320 kW
Hydraulic driven bag fans.
Computer controlled vent valves.
Bow fingers and three loop stern bag rubber skirts.

Hydrogenics is a Canadian-based company that develops and manufactures hydrogen generation and fuel cell products. The development of the company's fuel cells shows why it is only recently that hydrogen power trains have become a viable proposition. In 2001, its 25kW fuel cell weighed 290kg and had an efficiency that ranged between 38 and 45 per cent. The iLint's HD30 fuel cell has an output of 33kW and weighs 72kg with an efficiency ranging between 48-55%. Thus, since 2001, the company's fuel cells have become much more compact because of a fivefold increase in their energy density (from 86 to 458W/kg). Extrapolating this technology development to 2050 a target of another fivefold increase in energy density would bring the next generational technology to an energy density of 2500W/kg.[11]

Another example of fuel cell providers is SWISS Hydrogen. Their HyRex 100-E Range extender for heavy duty vehicle offers a 100 kW continuous power at only 95 kg. The life-time of this system is not optimum for this craft with its limit to 5,000 hours. However it is clear that there are fuel cell technology available that could be used. The weight is also significantly lighter than typical marine high-speed diesel engines, which typically comes up with 500 W/kg.

To get a benchmark of hydrogen supply needs it is useful to examine the German fleet of trains powered by hydrogen. It takes 15 minutes to refuel the iLint, which holds 178kg of hydrogen supplied at a pressure 350 bar. It consumes this at the rate of 0.3kg per kilometre. Thus, the fleet of 14 trains, covering, 600km/day, will require 2.5 tonnes of hydrogen per day. If this was produced by electrolysis, a wind farm of 10MW generating capacity would be required to power the required electrolysis plant with suitable back up. This, and sufficient hydrogen storage, will be required to ensure resilience of supply.

A typical HRS would require a 15MW power supply to produce the 280kg of green hydrogen per hour needed for 300 buses or 30 trains a day. It would consist of three 5MW electrolyzer modules with 30 bar feed pumps which feed six 30 to 300 bar compressors which pump the hydrogen into storage tanks holding up to 3,000 kilogrammes of hydrogen. When required, ten booster compressors raise its pressure from 300 to 450 bar and pump it into a 150kg buffer storage unit which feeds five dispensers and delivers the hydrogen at 350 bar.

Applying this data to the SES fleet operating from Ellis Island would require first to assess the fuel need. Hydrogen has larger energy density than diesel oil. The calorific value of hydrogen is 33 kW/kg vs. 11 kW/kg for diesel. The efficiency of a fuel cell is also better, where typically 50% can be achieved vs. 40% typically for a diesel combustion engine. This gives 61 g/kWh for hydrogen and 210 g/kWh for diesel. If the Ellis Island fleet consists of 5 vessels, each operating 2 hours at max speed a day and based on ESNA's Tern with the largest engine, the hydrogen consumption each day will be 5 x (2x1440kW + 2x85%x320kW) x 2 hours x 0.061

kg/kWh = 2,100 kg. This is similar to the German train example.

The current main challenge for hydrogen powered high-speed vessels are storage solutions for fuel onboard vessel. As hydrogen is heavily flammable precautions must be taken in positioning of the tanks and safety and fire systems. Further the tanks are very heavy. Current lightest solutions are pressure vessels made with carbon fibre material. An example of such supplier is Umoe Advanced Composites (www.uac.no). Typical storage pressure is 250 bar, with tank weight typically around 25 kg per kg of hydrogen.

Exterior form and platform development early Photoshop sketches are shown in Figure 18. Where the imposing stealth like angular form of the wheelhouse can be seen with the early forms of modular platform development. A more developed Photoshop sketch of the security response platform operating concealable miniguns is shown in Figure 19. The final CAD exterior is shown in front ¾ view and rear ¾ view in Figure 20 and Figure 21 respectively. The change in the proportions of the exterior form were a consequence of an inside out design approach, informed by a rigorous interrogation of user activities and human factor considerations. The user activities were developed through dialogue with end users and involved using scale models to develop SOPs and spatial awareness. The human factor considerations were achieved through an ergonomic evaluation of space and tasks using a CAD based ergonome.

The GA of the SES is shown in Figure 22, where the bow hinged ramp act as armour protection behind which there are two armoured personnel carriers. Aft of the APCs either side of the bridge are two autonomous armoured vehicles. Aft of the bridge there is a drone control centre on the port side and a crew cabin with 4 bunks on the starboard side. Aft of this a transverse corridor connects the troop deployment area on the outer port side to two inner changing rooms and a crew mess on the starboard outer side. Aft of this section is another transverse corridor which connects the port and starboard deck access areas, where security personnel can board RIBs in their deployed position. Fore of the central part of the corridor is a drone maintenance centre with the armoury on the port side of it and a store room on the starboard side. Aft of the centre of the transverse coridoor is a briefing/waiting area for security personnel. On the port side of this room is the medical facility, on the starboard side there is a storage room for security personnel equipment. At the transom of the vessel there are two RIBs that are secured transversely on brackets.

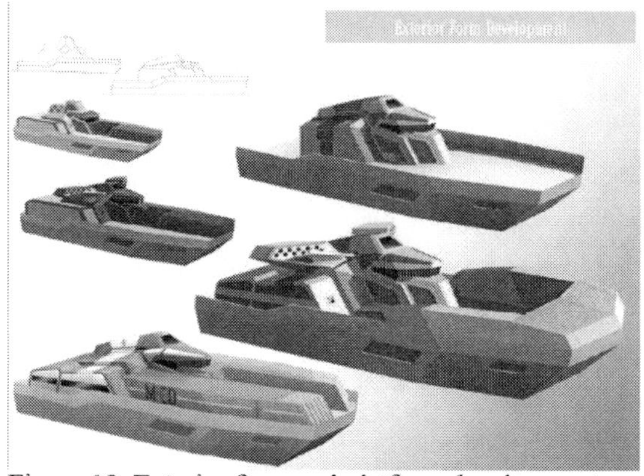

Figure 18: Exterior form and platform development

Figure 19: Exterior form and platform development

Figure 20: Exterior form and platform development

Figure 21: Exterior form and platform development

5. DEPLOYABLE SYSTEMS

The modular SES vessel can be configured with a range of vehicles and equipment. Whereby a range of platforms can be deployed from the vessels depending on the situation requirements.

RIB / PWC Deployment:

The platform is capable of deploying various types of small craft from RIBs to Personal Watercraft (PWC). The platform can rapidly deploy its organic craft to conduct numerous maritime operations from interdiction, insertion or recognisance. The ability to carry out mother daughter operations increases the effective area of operation (AOO). Deploying troops quickly and efficiently through close proximity to the troop preparation areas increases the platforms reaction times to rapidly changing situations. The platforms integrated C4ISTAR capability can update deployed craft giving the embarked personnel the ability to quickly alter their plan before insertion.

Figure 23: RIB deployment system

The RIB deployment system is shown in Figure 23, where the RIB are securely stored on brackets transversely on the transom are rotated through 90 degrees and lowered to deck level to facilitate boarding by security personnel.

APC Deployment:

The Armoured Personnel Carriers (APC) are rapidly deployable over the vessel versatile bow ramp system, shown in Figure 24. These vehicles are capable of deploying rapid reaction troops directly into the AOO to conduct numerous operations in support of the main mission. The vehicles give the platform the ability to influence further in land whilst still giving the troops a safe secure location to operate from. As with the maritime platforms the integrated C4ISTAR system is able to update the APC situational awareness whilst deployed. APC operations can include recognisance, find, fix or direct-action operations as well as humanitarian aid tasks.

Figure 24: Front ramp deployment

UAV Deployment:

The UAV deployment system located above the transom is shown in Figure 25. The platform drones are capable of both surveillance and direct-action tasks, deploying at extended distance in order to gather advanced intelligence prior to the arrival of the platform into the AOO. Once in the AOO the drones can be specifically tasked to counter specific threats. There are a total of 288 micro drones and 4 large modular drones. Micro drones can swarm areas to gather or track threats whist the larger drones can be directed to secure critical infrastructure or traffic hubs with either airborne or static mounted weapon positions. All drones feedback into the main platform C4ISTAR giving it a real-time picture vital when operating in such a built up congested environment. Direct targeting of assets is vital when operating in such a vast expanse utilising the vast amount of information supplied from all drones can provide the operational advantage required. The large modular UAV with combat module is shown in Figure 26, fitted with a M4A1 and infrared optical sight. The process of attaching the combat module to a building is shown in Figure 27. This enables the APCs to have a more effective engagement in the megacity environment through effective fire support.

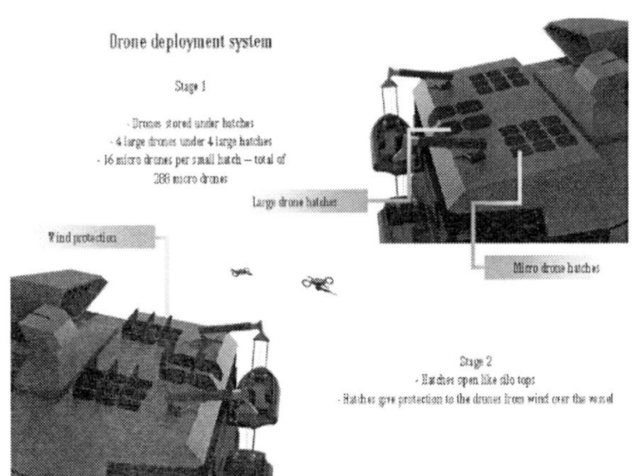

Figure 25: Location of UAV storage hatches

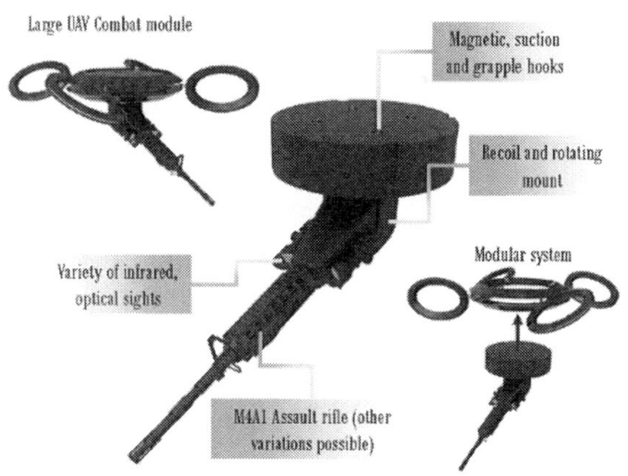

Figure 26: Large Modular UAV with combat module

Figure 27: Attaching combat module to building exterior

Figure 28: Modular UGV with riot shield

All drones are supported by the main platform, maintenance, power and servicing. This gives the platform a robot capability which can regenerate after prolonged deployment or maintain continual round-the-clock surveillance, security operations over a prolonged and high tempo mission. Where the hatches open forward to act as wind deflectors assisting the launch of UAVs while the vessel is underway at high speeds. The UGVs are also modular and primarily have a reconnaissance and fire support role for APCs. They can be configured as a riot shield for multiple security personnel, as shown in Figure 28. Where the implementation on non-leathal deterrents such as sonic weapons and tear gas can greatly enhance the capabilities of this platform.

6. CONCEPT OF OPERATION

The fleet of security patrol vessels operate using the Command and Control structure described in section X. In 2050 in order to optimise cognitive workload the HMI systems will consist of both audio and visual, with a focus on the use of augmented reality through the use of 3D hologram technology.

Figure 29: Bridge command centre

The bridge command centre is shown in Figure 29, where the navigator and pilot operate on the principles of DYNAV, with a weapons systems operator controlling the miniguns and missile system of the vessel. To the rear of the bridge is the command centre, where the command team operate through oversight of the theatre of operation projected as a 3D hologram augmented with mission critical information. This includes the deployment of security personnel in RIBs and APCs. The drone control centre shown in Figure 30, has 4 drone controllers and a drone commander for operational oversight. The drone controllers use a 360 degree augmented immersive reality projection system. The drone commander uses a holographic projection table to achieve oversight of the situational awareness provided by the drone operating in theatre of operation.

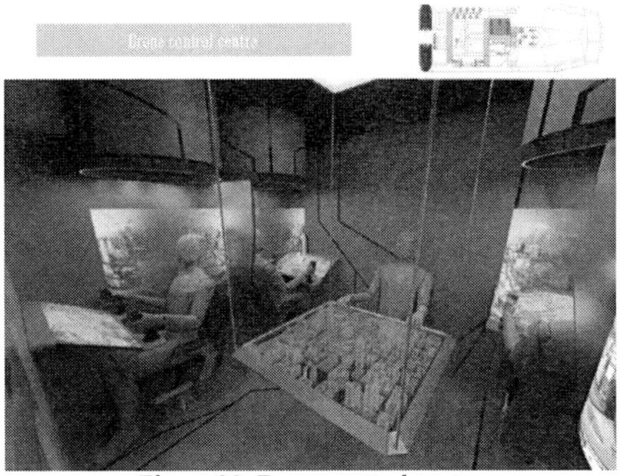

Figure 30: Drone control centre

7. CONCLUSIONS

The US Army Chief of Staff's Strategic Study Group (SSG) identified megacities as a security challenge within the future operating environment. The National Security Council's Global Trends 2030 reports that megacities present not only economic opportunity but also potential vulnerability due to their inherent need for security, energy and water conservation, resource distribution, waste management, disaster management, construction, and transportation. [4] The implication for the intelligence preparation for the urban battlefield (IPB) approach discussed is that the notion of distinct levels of analysis or slices of the intelligence challenge is a starting point, but that there are important connections and interactions among these levels that are complex and that sometimes create not only unexpected consequences but also, on occasion, surprising developments that can have strategic significance. While others are idiosyncratic. The deeper the under-standing of the city, the more likely it is that military units can act in ways that are aligned with the dynamics of the city, resulting in positive outcomes. [5] This would require, an ability to persistently monitor, collect, and interpret, in near real-time, the millions of bits of data associated with cell phone communications, social media postings, financial transactions, and the operational movements of these actors. The challenges associated with doing so will likely require the U.S. military to develop the ability to manage and interpret in a timely fashion the unending stream of data.

In a similar context the intelligence requirements for homeland security issues are critical. New York has a mixture of different cultures un-matched in the United States. With 600 separate cultural institutions, as well as over 51 million domestic and foreign tourists each year, agencies that provide services and security within the city must be able to understand and serve over 100 different nationalities often living in tightly knit Diaspora communities. Survey respondents [2] thought that the emphasis on the need for public surveillance would far outstrip privacy concerns, a conviction strongest in transitional and mature cities, where risk-based policing is most advanced. Technology is essential for risk-based policing, and a vital tool for general surveillance in cities wishing to increase police capacity.

Technology has been one of the drivers of transformation and is likewise driving the emergence of the new urban services paradigm. The top 10 technologies driving transformation are: Big Data; Internet of Things; Data Analytics Predictive & Prescriptive; Mobile device based sensing; Intelligent Transport; Open Data in Government; Smart Grid; Location & Condition Sensing Technologies; Citizen e-ID; Mobile Health Monitoring. While the use of technology varies across developing and developed countries, city administrators are increasingly looking at the following technologies to identify solutions to their urban challenges. [3] A key technology for 2050 will be the Distributed Common Ground Station (DCGS), which is envisioned to be a family of systems that provides multi-ISR processing and exploitation to the security task force. The DCGS Integration with the top 10 technologies driving transformation would ensure a closer coupling between C2 and intelligence, surveillance, and reconnaissance (ISR) processing systems.

The development of a command, control, communications, computers, intelligence, surveillance, and reconnaissance (C4ISR) architecture based on the Internet Protocol, a service-oriented architecture (SOA), and the developing capabilities of the Global Information Grid (GIG), promise to provide important elements of the flexible C4ISR needed for flexibly constituted fleet of SES security response vessels. By 2050 this will use Model-Driven Architecture (MDA), with fully established standards, to permit the rapid modification of applications software deployed across a variety of computing platforms. The objective of which as with the current DARPA JAGUAR is to transform the pace of operations to "the speed of thought." JAGUAR is embedding knowledge-based plan-development capabilities within a stochastic, dynamic control framework, to create a system that is self-aware, adaptable, and agile, and that scales to large problems and intricate domains.[10]

8. ACKNOWLEDGEMENTS

The authors wish to thank the following 3rd year Coventry University Boat Design students, whose professional collaborative EBDIG-IRC project constitutes the visualization work presented in this paper: Nicholas Morgan; Chengaw Li.

9. REFERENCES

1. HARRIS, M., DIXON, R., MELIN, N., HENDREX, D., RUSSO, R. and BAILEY, M., 'Megacities and the United States Army: preparing for a complex and

uncertain future ', Chief of Staff of the Army, Strategic Studies Group, June 2014, available at
2. MRC MCLEAN HAZEL and GLOBALSCAN, 'Megacity challenges: A stakeholder perspective', 2006, available at
3. World Economic Forum & PwC, 'Inspiring Future Cities& Urban Services: Shaping the Future of Urban Development & Services Initiative', April 2016 available at
4. KAUNE, P.N., 'Analysis of US army preparation for megacity operations', United States Army War College, April 2016, available at
5. WILLIAMS, P. and SELLE, W., 'Military contingencies in megacities and sub-megacities', U.S. Army War College Strategic Studies Institute, United States Army War College Press, Carlisle Barracks, PA, USA, December 2016, available at
6. GENTILE, G., JOHNSON D.E., SAUM-MANNING, L., COHEN, R.S., WILLIAMS, S., LEE, C., SHURKIN, M., ALLEN, B., SOLIMAN, S. and DOTY, J.L., 'Reimagining the Character of Urban Operations for the U.S. Army How the Past Can Inform the Present and Future',
7. https://www.porttechnology.org/technical_papers/analysing_threats_policies_and_solutions_in_port_security
8. IMO & ILO, 'Code of practice on security in ports', Tripartite Meeting of Experts on Security, Safety and Health in Ports, Geneva, 2003, available at ...
9. www.AISSecuritySolutions.com
10. National Research Council,'*C4ISR for Future Naval Strike Groups'*, 2006, Washington, DC: The National Academies Press. https://doi.org/10.17226/11605.
11. https://www.railengineer.uk/2018/01/05/hydrail-comes-of-age/

10. AUTHORS BIOGRAPHY

Dr S McCartan holds the current position of Course Tutor Transport Design, FAH, Coventry University. His key research area is TOI (Transfer of Innovation) from other sectors to the marine industry, in the areas of Design-Driven Innovation (DDI), advanced visualisation and Human Systems Integration (HSI). He leads the EBDIG (European Boat Design Innovation Group) network, which includes: Chalmers University; Genoa University; TU-Delft; and a number of leading European marine design consultancies.

Jon Hill holds the current position of Director at Trident Marine Ltd. He is responsible for providing consultancy support for system design, training and operational support. His previous experience includes holding the position of Warrant Officer within the Royal Marines, with extensive experience of maritime operations and training within the Landing Craft Branch.

Trygve H. Espeland holds the current position of Naval Architect and Co-Founder of ESNA. Who are a leading design consultancy specialising in the development of SES (Surface Effect Ships)

Dr Trevor Dobbins holds the current position of Director at STResearch Ltd. His experience includes maritime human factors work for the UK MOD, the RNLI and US Navy.

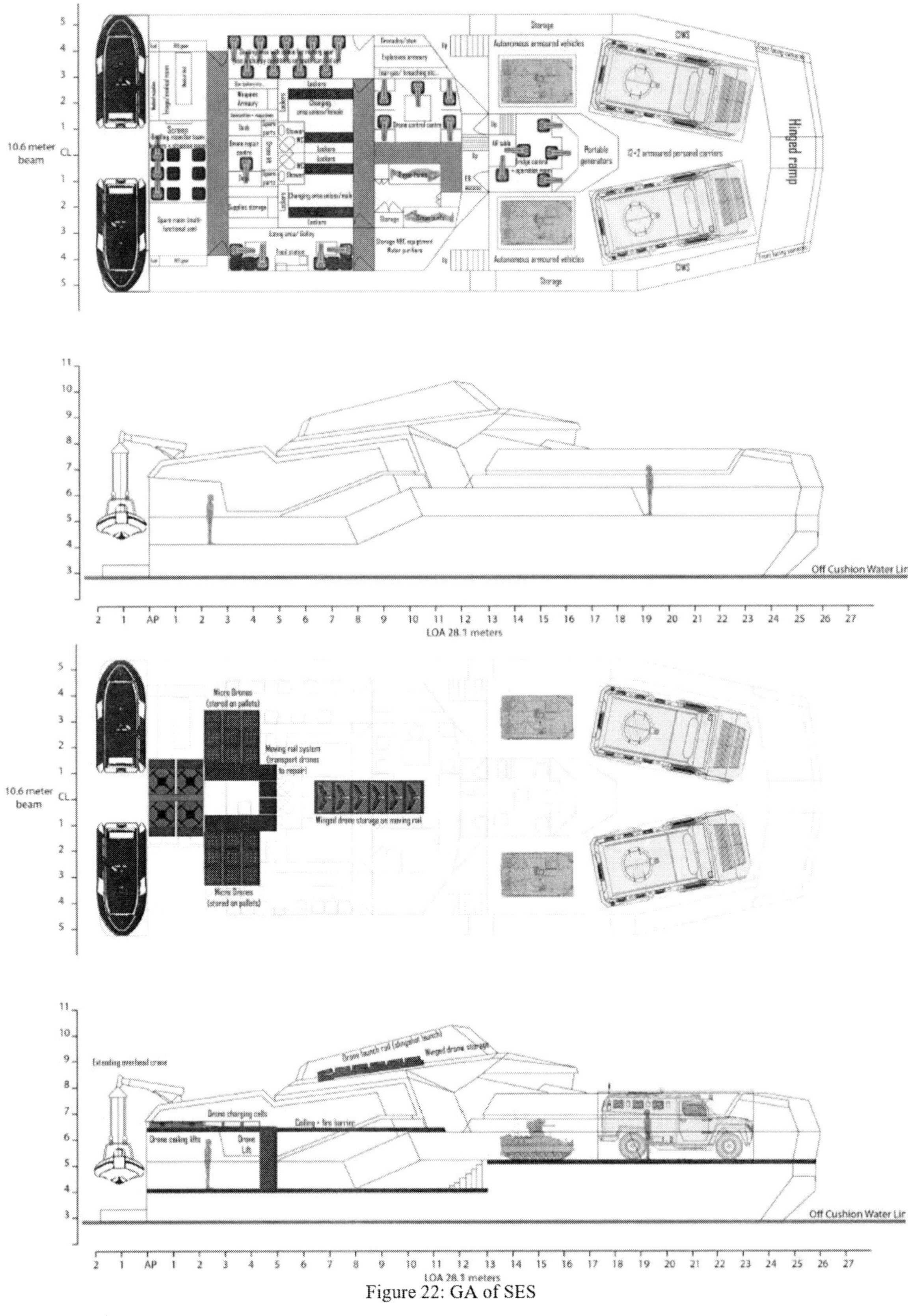

Figure 22: GA of SES

THE APPLICATION OF UNMANNED MARINE VEHICLES TO MARITIME SEARCH AND RESCUE

J. Dalziel and R. Pelot, Dalhousie University, Halifax, Canada

SUMMARY

The technology supporting Unmanned Vehicles, is rapidly improving. These vehicles are becoming more affordable and much more capable. Unmanned vehicles, land and air, are making inroads into Search and Rescue (SAR) response. Unmanned Marine Vehicles (UMVs) are already under evaluation for military applications, such as mine hunting and disposal, and surveillance, as well as scientific studies. Can UMVs play a significant role in Maritime SAR response? This paper looks into the issue from an Industrial Engineering perspective: Can they increase the safety of SAR personnel? Can they make SAR operations more effective (increasing the probability of saving lives)? Can they make SAR operations more efficient (a better use of limited resources)? Examples of the foregoing will be presented, applications suggested and the Cost-Benefit (benefit largely in terms of lives saved) discussed. The primary purpose of this Paper is to increase awareness of the possible life-saving benefits of this new technology.

NOMENCLATURE

AIS	Automatic Identification System
CAF	Canadian Armed Forces
CCG	Canadian Coast Guard
IMO	International Maritime Organization
JRCC	(Canadian) Joint Rescue Coordination Centre
Rescuees	Term used in this Paper for all persons being rescued, whether a Survivor or not
RHIB	Rigid hull inflatable boat, sometimes also referred to as a RIB
SAR Tech	Search and Rescue technician, a specially trained member of an elite group of primary care paramedics that provide on-scene medical aid and evacuation, may be deployed by parachute
TSB	(Canadian) Transportation Safety Board
UMV	Unmanned Marine Vehicle – Term used in this Paper for unmanned craft, whether remotely controlled or autonomous. Other terms, such as Unmanned Surface Vehicle (USV), Unmanned Underwater Vehicle (UUV), Maritime Autonomous Surface Ship (*MASS*) may be used. UMV is a very broad term and could potentially even involve 'marinized' Unmanned Aerial Vehicles (UAVs). All UMVs referred to in this Paper are surface craft.

1. INTRODUCTION

Since time immemorial humans have had beneficial interactions with the sea – for food, transport, and pleasure. However, these interactions have not always been without cost – the risk of death has always been present. Risks may be as varied as falling off a seaside cliff, being swept out to sea from a beach, or ship incidents, such as capsizing, foundering, grounding, collision, fire, man overboard. Reflecting the best in human nature, rescuers have responded to these incidents using boats, aircraft, helicopters, sometimes only a heaving line or entering the water themselves.

Up until the present time, rescue has required first-hand intervention by rescuers, potentially putting themselves at risk - sadly, on occasion, the rescuers have paid the ultimate price. In addition, particularly in cold waters, rescue may be very time dependent; the speed of the rescue response often making the difference between a rescue and a recovery (of a dead body). The rescue efforts may be delayed by heavy weather, and long distances (particularly in remote areas where response resources may be virtually non-existent, population sparse, and traffic densities low). Further complicating the issue, resources such as long-range aircraft may have restricted capability, limited to perhaps deploying life rafts and/or SAR Techs, or directing local response from vessels of opportunity, and the aircraft will have limited time on site. Recent improvements in the technology supporting Unmanned Marine Vehicles (UMVs) has the potential to significantly improve both the capability for Search and Rescue (SAR) response, and the safety of rescue personnel.

This paper looks at the current, and near-term future prospects for utilizing Unmanned Marine Vehicles for Search and Rescue from an Industrial Engineering perspective. Industrial Engineers design solutions to improve the performance of complex systems of people, technology, and information. An overview will be given of the historical context, of the technology commercially available for UMVs, and of some recent studies into their application for SAR. However, the primary focus will be addressing questions such as:

- Do Unmanned Marine Vehicles (UMVs) have a role in maritime Search and Rescue (SAR) systems, and if so, what roles might they fulfill?
- How would UMVs be integrated into a comprehensive SAR system?
- What hurdles to acceptance should be anticipated? What are the reasonable expectations of these craft?

Some examples of the circumstances, places and applications, where UMVs could make a positive contribution, will be given. Finally, recommendations will be made as to the 'Way Forward'. A key intent of this Paper is to help raise awareness of the potential contribution of Unmanned Marine Vehicles to Search and Rescue.

2. SEARCH AND RESCUE – HISTORICAL CONTEXT

"a large body of men ... in constant readiness to risk their own lives for the preservation of those whom they have never known or seen, perhaps of another nation, merely because they are fellow creatures in extreme peril." (Sir William Hillary, founder of the Royal National Lifeboat Institution (RNLI), 1823) [1]

When Sir William Hillary wrote those words 200 years ago, there was little organized, formal response to marine disasters. Over the past two centuries improvements in rescue response, technology, safety standards, enforcement of regulations and education have reduced risks for both mariners and for those who seek pleasure on the water. Two hundred years ago when a ship was wrecked near a harbour entrance, often the ship's crew would take to the rigging for survival and the townsfolk of coastal communities would risk their own lives by putting to sea in whatever craft was available. The large number of tragedies eventually led to the first purpose-designed lifeboat in England in 1789, the 'The Original'. However, significant loss of life at sea continued, leading to the establishment of the Royal National Lifeboat Institution (RNLI) in 1824. Similar organizations were founded in many other countries around the same period, and world-wide growth of the maritime SAR organizations has continued to the present day, as evidenced by the widespread membership in the International Maritime Rescue Federation. In an effort to formalize and improve ship safety and maritime rescue response, the International Convention for Safety of Life at Sea (SOLAS) was introduced, largely as a result of the Titanic disaster. SOLAS Chapter V Regulation 7 states:

"Each Contracting Government undertakes to ensure that necessary arrangements are made for distress communication and co-ordination in their area of responsibility and for the rescue of persons in distress at sea around its coasts. These arrangements shall include the establishment, operation and maintenance of such search and rescue facilities as are deemed practicable and necessary, having regard to the density of the seagoing traffic and the navigational dangers and shall, so far as possible, provide adequate means of locating and rescuing such persons" [2].

The International Maritime Organization has continued to update SOLAS to improve safety. In addition, the International Convention on Maritime Search and Rescue (the SAR Convention), which entered into force in 1985, developed an international SAR plan, so that the rescue of persons in distress anywhere at sea will be coordinated by a SAR organization and, when necessary, by cooperation between neighbouring SAR organizations. Distress communications also improved, from visual sighting of distress flares through radio distress communications to the Global Maritime Distress and Safety System (GMDSS) and Emergency Position Indicating Radio Beacons (EPIRBs) of today.

Up to around the end of the 19th century, lifeboats were propelled by oar and/or by sail. Early in the 20th century, engine powered lifeboats started to be introduced. These were generally displacement craft limited to their hull speed, about 8 knots. Modern high-speed lifeboats have speeds of 25 knots or more, extended range, and extensive electronics for navigation, communication and searching. Aircraft and helicopters have contributed significantly to rescue response. As safety technologies have advanced, the training of SAR personnel has increased; for instance, SAR Techs are able to provide on-scene medical care and are often deployed from aircraft or helicopters. Modern communications, lifeboats and aircraft now operate as an integrated response system for maritime emergencies. In the next few years Unmanned Marine Vehicles will likely also be integrated into the emergency response system.

3. UNMANNED MARINE VEHICLE TECHNOLOGY

3.1 OVERVIEW OF TECHNOLOGY

The technology for the control of Unmanned Marine Vehicles is well developed in a large number of countries and widely available on a commercial basis.

3.1 (a) Control & Communication Technology

Most UMVs are controlled using VHF/UHF radio frequencies, broadband radio, or by satellite-based control systems. The radio-based systems are line of sight, the range limited by power and antenna height. Generally, this will not exceed 15 nautical miles. Some broadband radio systems can provide communication with significant data capacity for up to 300 km (180 nm), depending on antenna height. With satellite communication the distance can, in theory, be infinite. The control station could be on a nearby or distant vessel, a nearby or distant aircraft, or a fixed or mobile station ashore, depending on the communication regimes chosen. Communication could include visual support from overflying aircraft or UAVs, integrate several manned and unmanned craft, split between a variety of control stations - all depending on the capabilities and the complexity of the systems chosen. [Figure 1] illustrates these concepts [3, 4].

Figure 1: Maritime Broadband Radio (Kongsberg) [3].

Figure 2: UMV Control Diagrammatic (ASV) [44].

The UMV would be equipped with one or more cameras, optical, low light, infrared, sound detection, sonar, etc., providing a first-person view capability back to the operator, who can evaluate the situation on scene without the hazards of physically being there [Figure 2]. In addition it could also be equipped so that the operator can communicate with the persons being assisted. In many cases, the UMV may also be equipped with various navigation capabilities, such as return to home (if control communication is lost), collision avoidance technology, etc.

3.1 (b) Assisting the Rescuees

In cold water, people may quickly lose control of their arms and legs and become unable to help themselves. How can a UMV assist someone who is unable to assist themselves? This may be the area that requires the most detailed technical consideration. The solution, if there is one, may be very situation specific. For instance, as noted in the ICARUS study, the 'AGaPaS' has a specially designed pick-up device. Another device is a rescue capsule, which has an inflatable life ring [3].

3.1 (c) Other Technical Considerations

A full concept of operations would span a large range of elements including the following. Power may be by diesel, gasoline, or electric (batteries). For the SAR service in most regions, the UMV would have to be very robust, also with adequate stability. Launching and retrieval arrangements for the UMV would depend on the mode of operation. Appropriate level of protection from the elements for the rescuees could be incorporated.

3.1 (d) Regulatory Regime

The regulatory regime for unmanned and for autonomous vessels is in a state of transition, largely due to the rapid rate of change and advancement of these vessels, both technically and in their application. The IMO has started a scoping exercise to determine what change might be needed to support the use of maritime autonomous surface vessels. Several regulatory 'recommendations' currently exist – such as the Lloyd's Register (LR) Code for Unmanned Marine Systems (February 2017) and the UK Maritime and Coastguard Agency (UK MCA) Voluntary Code for 'Maritime Autonomous Surface Ships' (November 2017). In the absence of other regulations, these Codes provide a regulatory framework. Highlighting a few points, the LR Code is stated to be applicable to non-Convention ships (in general less than 500 tons Gross Tonnage) and specifies a range of levels of autonomy – from full remote control with no onboard decision making, to fully autonomous decision making. The MCA Code is for vessels up to 24 M in length, with a range of classes depending on size and speed, and the certification of operator training, competence and watchkeeping [5, 6].

In addition to compliance with the regulatory regime, the operator should take steps to ensure the UMV will not pose a risk to other persons and vessels, the potential rescuees, and the environment. The risks should be no more than those of a conventional vessel.

3.2 STUDIES AND APPLICATIONS TO DATE

The applications of UMVs to date have generally been military, particularly for aspects such as mine detection, where their small size and low acoustic signature make them safer in a minefield, and should they be destroyed, there will not be loss of life. Another common application is in oceanographic work, where autonomous craft, such as sea gliders, have been used for extended oceanographic data gathering. However, the range of uses for unmanned/ autonomous craft is rapidly increasing, due in particular to both cost and safety concerns.

The applications of UMVs to Rescue, or Search and Rescue, is progressing more slowly; some very small craft, are in use, or are proposed, such as the Emily, U-Safe, AGaPaS, primarily for beach rescue, and man overboard response.

3.2 (a) ICARUS

The EU-funded ICARUS (Integrated Components for Assisted Rescue and Unmanned Search operations) research project started in 2012, aiming to develop robotic tools which can assist the "human" operator during SAR operations; 24 private and public partners from 10 nations participated. The ICARUS project sought to bridge the gap between the vast amount of literature in the Research community and end-users (SAR organizations), by developing a toolbox of integrated components for unmanned Search and Rescue. In July 2015, ICARUS project partners simulated a maritime crisis management scenario at a Naval Base in Portugal. The demonstration showcased how Unmanned Surface Vehicles (USVs) and Unmanned Aerial Vehicles (UAVs) – linked together in a shared robotic network - can assist "human" search and rescue teams in maritime incidents, such as the sinking of a large ferry boat or a cruise ship. The ICARUS project included a 7 M RHIB (the U-Ranger), a 4 M catamaran and a 1.45 M unmanned battery powered, water-jet driven, rescue capsule equipped with an inflatable rescue ring, which could be deployed from either of the larger craft [3].

3.2 (b) Naval Applications - Unmanned Warrior (2016)

Unmanned/autonomous craft, underwater and surface, have been evaluated and employed by various navies for some years, largely for tasks such as guarding against mines and other hazards, and allowing surveillance to take place without requiring a manned vessel to enter a hazard zone. In 2016 the Royal Navy with its partnerships in industry, academia and allied nations, hosted a real life, large scale demonstration of unmanned and autonomous systems codenamed "Unmanned Warrior". Some of the questions were: Can they work together? Can they provide operational commanders with a better understanding of their environment? Do they provide a competitive edge? Can they reduce the risks faced by our armed forces? Can they perform in the real world? The exercise allowed the command structure of the various navies to interface and evaluate a variety of technologies providing commanders at sea a means to identify autonomously vessels of interest, improving their decision capability. Unmanned Warrior was about: Integrating systems and bringing them into a real-life operation situation and getting the data into the hands of an operational commander; Whether there is potential of getting the new technologies into service more quickly than expected; Creating operational advantage for the Royal Navy (and allied navies). From a Canadian perspective, Canada's participation in the exercise and the knowledge and experience passed on to the Canadian Armed Forces (CAF) informs them on the state-of-the-art possibilities and helps inform their eventual decisions on procurement, training, concepts of operation, and the way-ahead for the role of unmanned vehicles. Unmanned Warrior 2016 appears to have increased an understanding of the operational advantages and limitations of unmanned vessels, their interoperability between modes (air, surface and subsurface), between navies, and between navies and their suppliers [7, 8].

3.3 REPRESENTATIVE EXAMPLES OF UNMANNED MARINE VEHICLES

The following is a small sampling of the (surface) UMVs that are currently available, or in various states of concept, design and construction.

3.3 (a) Small / Mid-Size Unmanned Marine Vehicles

ASV Global - UK - Builds a range of small to mid-size unmanned / autonomous craft, primarily for military and oceanographic applications. It also converts existing craft for unmanned operation, using its ASView system such as outboard powered Rigid Hull Inflatable Boats. ASV and its craft participated in the Unmanned Warrior exercise in 2016. As an example, ASV Global integrated its proprietary control system ASView onto the BAE Systems Pacific 950 RHIB to turn the standard vessel into a craft that has the ability to operate unmanned, autonomously and by direct remote control. The modified P950 is capable of operating autonomously for up to 12 hours at a time on either a pre-planned route or via remote control. It can operate up to 40 km away from its parent ship and reach speeds in excess of 38 knots (44 miles per hour), providing unique ship-launched maneuverability and enhanced situational awareness to support the decision-making of its operators [Figure 3]. Another example is the conversion in 2017 to unmanned capability of an outboard powered Canadian hydrographic launch. ASView enables the launch to follow planned missions giving it the capability to operate as a standalone survey launch controlled from the shore or as a force multiplier controlled from another vessel. ASView seamlessly interfaced with the launch's existing engine, steering and navigation systems to enable autonomous operation. The launch is connected to a remote station via a suite of IP radios enabling real-time monitoring of the survey acquisition data and vessel parameters. The launch is equipped with an advanced collision avoidance system using radar and AIS to ensure its safe operation. Situational awareness is provided by five onboard cameras with audio feedback. ASView monitors the launch's onboard sensors including depth, engine and battery health status alerting the remote station of any potential hazards. Following on-water testing, ASV Global delivered its (UK) MCA recognised Maritime Autonomous System Operator training course to the Canadian operators [9, 10].

Figure 3: RHIB converted for Unmanned Operation (ASV) [44].

Marine Applied Physics Corporation (MAPC) - USA - MAPC developed several unmanned craft, including manned and unmanned versions of the GARC (Greenough Advanced Rescue Craft) [Figure 4]. The GARC is a 12 to 14 ft. personal watercraft type rescue boat, designed for durability in breaking surf and high waves. According to MAPC in the manned configuration, the GARC is a high-performance rescue vehicle. It can be deployed from C-130 and C-17 aircraft and is in use by USAF units worldwide. The vessel is used to rescue downed USAF air crews and passengers on commercial airliners that ditch in the open ocean. The Para Rescue personnel that jump with these vehicles are equipped to provide medical care, food, water, and shelter to survivors for up to 3 days while waiting for ships to arrive on scene. A total of 32 craft have been built and are in service. The GARC has been mission-tested in a typhoon that produced 30-foot seas. The GARCs proved their capability by rescuing Japanese Coast Guard teams on personal watercraft and remained operational in Sea-State 7 for 3 days (until the mission was complete). Unmanned variants of the GARC rescue craft have applications in the U.S. Navy and Special Warfare communities as well as in various commercial offshore applications. MAPC is currently under contract to employ unmanned GARCs in cable survey operations for off-shore wind projects and other federal government sponsors. MAPC also developed the TALON kite electronics system employed by the DARPA 'Sea Hunter' to give an extended range of view (at 1,000 ft height, over 60 km range), a smaller version is under test for the GARC [11].

Figure 4: GARC- Greenough Advanced Rescue Craft (MAPC).

3.3 (b) Large Unmanned Marine Vehicles

DARPA (Defence Advanced Research Projects Agency) - USA - The 'Sea Hunter' [Figure 5] is a 130ft unmanned trimaran, designed as the forerunner of a new class of oceangoing vessels which can travel thousands of kilometers over open seas for months at a time without crewmembers, but always with remote human supervision. Potential missions initially will include submarine tracking and countermine activities, estimated daily operating costs are about $15,000 to $20,000, (in contrast to the daily costs of $600,000 cost to operate a manned helicopter, or $700,000 to operate a destroyer) [12].

Figure 5: DARPA 'Sea Hunter' with TALONS Kite (DARPA) [45].

Kongsberg - Norway - Kongsberg is involved in both the control/communications systems of unmanned/autonomous vessels, and several associated vessel projects. One project is the 120 TEU, 3200 ton, electric propelled cargo vessel 'YARA Birkeland' for a specific service in north Norway, scheduled for delivery in 2019. Another proposal, in conjunction with Vancouver based naval architects Robert Allan Ltd, is for an autonomous 20 M fireboat, the 'RALamander 2000'. The K-Mate system is the common autonomous control engine to be used by unmanned and autonomous vessels delivered by Kongsberg Maritime [13, 14].

Rolls Royce - Finland, UK - In the summer of 2017 Rolls Royce demonstrated the unmanned operation of a commercial tug in Copenhagen. It is also working on various projects to promote autonomous shipping, including commercial and naval vessels. Earlier this year Rolls Royce announced the opening of its new R&D Centre in Finland, enabling Rolls-Royce and its partners to carry out projects focused on autonomous navigation, the development of land-based control centres, and the use of artificial intelligence in future remote and autonomous shipping operations [15, 16].

3.3 (c) Some Unmanned Marine Vehicles Specifically Intended for Rescue

AGaPaS (Autonomous Galileo-supported Person rescue at Sea). - Germany - A system proposed by University of Rostock researchers for a satellite-based system to respond automatically to man overboard cases from ships and offshore platforms. The vessel is a self-righting catamaran, approximately 4 M by 2 M, with a rescue platform which can scoop persons out of the water [3].

EMILY (Emergency Integrated Lifesaving Lanyard) - USA - is a small remotely operated autonomous buoy that aims to assist the lifeguards at crowded beaches, providing them a safe and fast response means. It has also been used to assist the rescue of migrants in the Mediterranean [17].

U-Safe - Portugal - A waterjet propelled horse-collar type life ring which has a control range of about 3 NM [18].

4. CONTRIBUTION OF UNMANNED MARINE VEHICLES TO SEARCH AND RESCUE

In what ways do Unmanned Marine Vehicles (UMVs) have a role in maritime search and rescue (SAR)? What needs do they fulfill?

- Can they increase the safety of SAR personnel?
- Can they make SAR operations more effective (increasing the probability of saving life)?
- Can they make SAR operations more efficient (a better use of limited resources)?

How can UMVs best be integrated into a comprehensive SAR system? What are the barriers to their successful introduction, to maximize efficiency and effectiveness, while at the same time increasing the safety of SAR personnel?

Some of these issues can be analyzed intuitively by 'front-line' SAR personnel - Lifeboat crews, SAR Co-ordinators, operational and technical staff, once they understand the performance capabilities and limitations of UMVs. From an Industrial Engineering perspective, utilizing an understanding of the needs, environment and technology, it is possible to propose solutions to improve the performance of these complex systems of people, technology, and information. This will give management additional assistance in evaluating the possible introduction of UMVs.

In the following we provide a broad overview of answers to the foregoing questions, and some of the applications where UMVs could make a positive contribution to SAR response. As an aid to management a Cost-Benefit Analysis can be used as a preliminary tool to evaluate the case for the introduction of Unmanned Marine Vehicles to a Search and Rescue system. The costs, although not inconsequential, may be very small in comparison to the benefits, which include reducing financial costs of using traditional rescue assets, and more importantly lessening the human costs, in terms of suffering and loss of life. In the following sections, the Needs, the Capabilities, and the 'Benefits' will be considered.

4.1 SAFETY OF SAR PERSONNEL – THE NEED

Up until the present time rescue has required first-hand intervention by rescuers, potentially putting them at risk - sadly, on occasion, the rescuers have paid the ultimate price. The following cases represent some of the dangerous situations which may be encountered by SAR personnel.

"Canadian Coast Guard Chief Officer Gregory Paul Peddle, S.C., of Spaniard's Bay, Newfoundland and Labrador. Chief Officer Peddle and his colleagues, Senior Engineer Pierre Gallien and Leading Seaman Raymond C. Welcher, lost their lives on October 15, 1989, when their fast rescue craft overturned in an attempt to rescue a diver off Middle Cove, Newfoundland. Chief Officer Peddle was awarded the Star of Courage" [19].

Chief Officer Peddle and his shipmates were lost on their Fast Rescue Craft, launched from a Coast Guard ship, as they attempted to rescue a sports diver in near shore, breaking seas, conditions.

Loss of crew (4) and rescued fishermen (3) from German Sea Rescue Cruiser 'Adolph Bermphol' & daughter boat 'Vegesack', 23 February 1967. Daughter boat was launched with 2 crew, leaving 2 crew on mother vessel. Three Dutch fishermen were rescued. In hurricane force winds, local navigation marks were extinguished. While attempting to recover the daughter boat, the two crew on mother vessel were swept overboard; subsequently all persons were lost [20].

The case of the Adolph Bermphol highlights some of the risks in launching and recovering a rescue boat, particularly in very heavy weather, and near shore conditions.

Caitriona Lucas, who was a mother of two children, died after a rescue boat capsized and threw the three people on board into the sea off Kilkee, Co Clare, yesterday. At approximately 12.11 hrs, on the second tasking of the day and whilst searching a cove to the east of Foohagh Point, the Delta RIB capsized throwing all three crewmembers into the water [21, 22].

An Irish Coast Guard volunteer lost her life in 2016 while participating in a near shore search for a missing man, believed to have fallen over a cliff.

Near shore rescues, particularly in breaking sea and surf conditions, are very hazardous.

"Sergeant Gilbert, a search and rescue technician was taking part in a rescue operation near Igloolik (Hall Beach), Nunavut. Unfortunately, in the course of this dangerous mission, he lost his life. The operation itself resulted in the rescue of two local citizens" [23].

Sergeant Gilbert was one of several SAR Techs who parachuted from a Hercules into the sea, near Igloolik, to assist walrus hunters stranded in a small open boat. The Herc. departed the scene due to low fuel, about seven hours later (after a 14-hour, 2,200 km flight) a rescue helicopter arrived and was able to pluck the survivors, and Sergeant Gilbert's body, from the sea [Figure 6] [24].

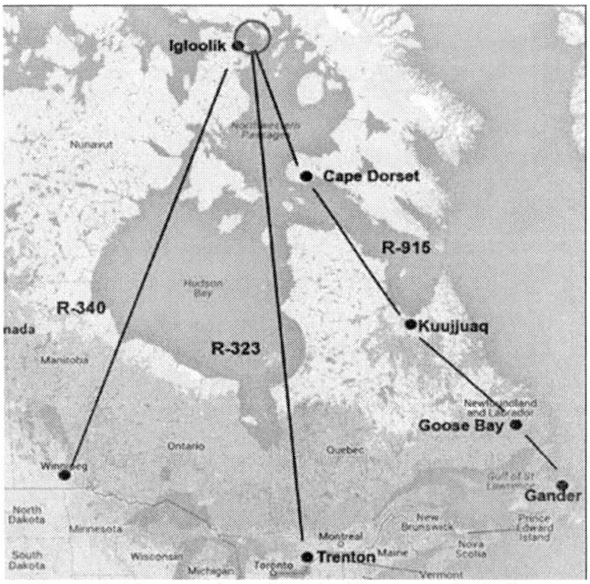

Figure 6: Air operations for Igloolik rescue, including helicopter response from Gander, Newfoundland [24].

"the ARKONA had detected and reported highly toxic gases in the atmosphere surrounding the LISCO GLORIA's bow and portside area, after having carried out a measuring procedure. The ARKONA therefore retreated from the polluted area" [25].

As noted above, another situation potentially dangerous to rescue personnel is responding to a fire, where toxic gases may be present.

And finally, the RNLB 'Solomon Browne' a 47 ft. Watson Class lifeboat was lost with all hands during an attempt to rescue the remaining crew of the small cargo ship 'Union Star' in horrific near shore conditions,

Lt Cdr Smith USN, the Sea King pilot, later reported that: *'The greatest act of courage that I have ever seen, and am ever likely to see, was the penultimate courage and dedication shown by the Penlee crew when it manoeuvred back alongside the casualty in over 60 ft breakers and rescued four people shortly after the Penlee had been bashed on top of the casualty's hatch covers. They were truly the bravest eight men I've ever seen, who were also totally dedicated to upholding the highest standards of the RNLI' [26].*

4.1 (a) Safety of SAR personnel – Unmanned Marine Vehicle Contribution

UMVs could potentially replace humans in some of these front-line activities, particularly in the most dangerous situations. Many of the foregoing are near shore rescues, where the rescuers may be subject to extreme breaking waves, and unforgiving obstacles, such as rocks, cliffs and shorelines, and parts of ships. For inshore rescue, there are many operational scenarios, such as:

- Bring the rescuees to a safer location for rescue by a manned craft;
- Establish a connection, such as a line, between the rescuers and the rescuees;
- Close up evaluation of the on-scene conditions, including determining if it is 'safe' to attempt a rescue;
- Acting as a secondary response boat for rescuers who enter hazardous areas, such as surf zones, perhaps with a line attached to the rescuer / rescuer's boat.

UMVs could be deployed from aircraft to accompany SAR Techs who parachute or are lowered into the sea.

- Be manoeuvred to SAR Techs and rescuees;
- Act as place of refuge;
- Act as communication hub, particularly if over-flying aircraft need to depart for fuel;
- Provide radar / AIS target for ships / aircraft.

In most cases, these UMVs would be relatively small – such as the personal water craft style UMVs, like the GARC (USA), RescueRunner (Sweden, if fully converted for remote control operation), and various inflatables / RHIBs (if converted for remote control operation). For firefighting in hazardous and in toxic environments UMVs could be the front-line fire boats, if fitted with water monitors.

4.2 CAN UMVs MAKE SAR OPERATIONS MORE EFFECTIVE?

Effectiveness in SAR can be defined as the degree to which something is successful in producing the desired result, i.e., saving life. Two examples of effectiveness are examined below.

4.2 (a) Effectiveness - Recovery of Rescuees by a Ship – The Need

Retrieving people from life rafts and small boats, or directly from the water, can be a very difficult operation for potential rescue vessels to perform, and it may also be very dangerous for the persons being rescued. In many, if not most, cases the rescue vessel may seek to launch a rescue boat, to assist with the retrieval of personnel, and marshal life rafts, etc.
There are statutory requirements for specified vessels to be equipped with rescue / emergency boats to assist with rescuing persons in distress. In Canada, these are in the Life Saving Equipment Regulations, made pursuant to the Canada Shipping Act (2001) [27]:

- 2 (1) Rescue boat means a vessel designed to be used for rescuing persons in distress and marshalling survival craft.
- 39 (b) A marine anti-exposure work suit for each member of the rescue boat crew;
- 148 (1) Rescue boats and emergency boats shall be stowed in a state of continuous readiness and capable of being launched in less than five minutes;

Notwithstanding the statutory requirement to be capable of being launched in less than five minutes, many Masters consider that at least twice that length of time (10 minutes) may be necessary due to the length of time required for the crew to don anti-exposure gear. An additional factor is the Master's responsibility for the safety of his own crew - the decision to launch may be affected by the risks to his crew, such as weather, the perceived condition of the rescuees, and the back-up should things go wrong.

"Search and rescue is always a fine balance between courageous intervention and managed risk." As HMCS Shawinigan departed the Arctic…boat crews struggled in reduced visibility and surf attempting to rescue freezing hunters…the Captain made a heartfelt but correct decision to suspend further attempts. The probability of adding more casualties to the emergency could not be risked [28].

In this case, the Master determined that, considering the circumstances of the persons in danger, the weather and sea conditions, the nature of his ship's rescue boat and launch / retrieval equipment, and the lack of backup resources should things go wrong, the rescue attempt needed to be postponed. The ship stood by, and fortunately the hunters were rescued by helicopter the following day.

The Seaforth Highlander spotted flares from a lifeboat and headed towards it. A lifeboat manoeuvred alongside - the Seaforth Highlander was able to get a line attached to the lifeboat…but the heavy seas snapped the line and capsized the lifeboat, and ultimately the Seaforth Highlander was unable to rescue a single man [29].

When the Oilrig 'Ocean Ranger' was lost with all hands (84 persons) in February 1982 in a ferocious storm off the coast of Newfoundland, it is reported over 50 of her crew made it into lifeboats, or directly into the water. Although three supply vessels were standing by, attempting rescue, all personnel were lost. The vessels standing by attempted to rescue those from the lifeboats and also attempted the use of life rings to assist persons in the water, all without avail. The USCG report recommended consideration of other devices for rescue from the water, such as nets or baskets.

After some time, the ISABELLA opened a port in the hull and released a slide. The people in the water and a couple of crew members with survival suits were helped up onto the slide by other crew members. Sixteen people from this raft were rescued this way and one by helicopter. The lifesaving equipment of vessels participating in the rescue operation proved unsuitable for rescuing people from the water in the prevailing heavy weather conditions. Systems should be developed for enhancing the ability of passenger ferries to rescue people from the sea in heavy weather [30].

When the ferry 'Estonia' sank in the Baltic in in September 1994, 989 persons were onboard, only 138 were rescued. Many went down trapped inside the ship, however over 300 were reported to have made it to the open decks and could potentially have been rescued. Despite nearly 30 ships eventually participating in the rescue, most survivors were rescued by helicopter, and only 34 of the survivors were rescued directly by ship (of these 34, 16 pulled up the rescue slide which had been deployed by the ferry 'Isabella').

By about nine or 10 p.m., the MV Arctic arrived on the scene. It was able to get a line hooked on to the life raft and hold it alongside, but because of the rolling of the big ore carrier, the fishermen decided it was too dangerous to climb the ship's side. About another hour later, the (trawler) Paamiut arrived. It had a crane and the proper boats to launch given this kind of weather. Two at a time, they were able to bring back the fishermen [31].

After the fishing vessel 'Atlantic Charger' sank in September 2015, north-east of Labrador, the crew members spent hours in a life raft until a suitable rescue ship arrived, despite rescue aircraft flying overhead, and a cargo ship attempting to offer assistance [Figure 7].

Figure 7: Atlantic Charger sinking (CAF) [37].

It may be unsafe – or simply impossible – to bring survivors alongside your ship directly. You may have to find another way of reaching them. One way to do this is to launch a rescue craft, if this can be achieved safely. Another way is to pass a line. Launching a rescue craft will serve three purposes: .1 it will make the final approach to the recovery object easier; .2 primary recovery (into the rescue craft) will be easier, because of the rescue craft's lower freeboard and similar motion to that of the recovery object; and .3 completing the recovery by returning to the ship and being lifted back aboard using the rescue craft's own recovery system should also be easier. Ultimately, however, the only option may be to abandon the attempt at recovery and to stand by, supplying whatever assistance you can until a more capable recovery unit arrives or conditions ease [32].

The foregoing examples are summed up by IMO MSC.1/Circ.1182 'Guide to Recovery Techniques' pointing out the difficulties which may be experienced trying to save people from lifeboats, life rafts or directly from the water, even when rescue seems at hand.

4.2 (a) (i) Recovery of Rescuees by a Ship – Unmanned Marine Vehicle Contribution

UMVs can increase the effectiveness of personnel recovery from the water in several ways:

- Immediate launch without having to wait for crew to don anti-exposure gear;
- Launch in heavy weather without endangering crew members;
- Proceed into waters dangerous for a manned rescue craft;
- Marshal life rafts alongside rescue vessel;
- Provide temporary shelter in cases where it may not be possible to immediately safely recover the personnel from the water, including providing continued location and communications with the rescuees;
- Provide a backup resource for SAR personnel who have been launched in manned rescue craft;
- Be dropped from an aircraft, and rendezvous with rescuees and the rescue ship.

4.2 (b) Effectiveness - Widening the Search Pattern – the Need

Three missing kayakers from Pennsylvania were located in Orleans County Tuesday morning. One was found dead. The two men drifted, likely more than 15 miles, and for several hours. The US Coast Guard dispatched two rescue boats — one from Youngstown and another from Rochester — after being notified of their disappearance around 7 p.m. As night fell and severe storms blew in, the men told authorities they got as close as 50 yards from what they thought was a rescue boat. But the boat could not hear them, and they continued to drift. Boats came in due to severe weather at 9:47(PM). Joe and Nathaniel continued to drift, likely a few miles east of Point Breeze until they could make it to shore. They walked for about an hour before coming upon a jogger early Tuesday morning. "At some point, one of those rescue vessels was close to those individuals that survived, but they just couldn't hear them. One of the men had lost their voice," Bourke said. Jacob's body was found by coast guard about two miles east of Point Breeze, and about a half mile from shore [33].

Three kayakers went missing in Lake Ontario in September 2017, after encountering problems in worsening conditions. Despite search efforts, including boats, aircraft and ground search, they were not found until the next morning, meantime one had perished.

4.2 (b) (i) Widening the Search Pattern – Unmanned Marine Vehicle Contribution

One of the issues in searches for missing persons is the breadth of the effective search pattern. If it is a shoreline search for a missing kayaker, or a search at sea for a person lost overboard, the effective visual range of sight for small non-radar reflective targets can be limited, the observer's height above water (generally less in a SAR cutter than in a ship), the sea conditions, the visibility (mist, fog, daylight, dusk or night), the clothing of the rescuee, concern about navigational hazards, the attentiveness of the observers – these are all factors. In addition, particularly for shoreline searches, manned vessels must often remain a distance offshore due to navigational hazards. For instance, one or more UMVs could serve to broaden the search track of a SAR cutter. The UMV could also be used nearer to shore, in areas of rocks and other navigational hazards. For an offshore search, a group of (say 6) UMVs could accompany the mother ship, providing a very wide search track, with perhaps some overlap to reduce the possibility of missing the person/persons. The UMV could also be equipped with an aerial device (tethered drone, balloon, kite), the breadth of search would then be wider. (The aerial device could also include communication link, thereby increasing the control range of the UMV).

4.3 CAN UMVs MAKE SAR OPERATIONS MORE EFFICIENT?

How can efficiency be defined: Fewer resources for the same outcome? A better outcome for the same resources? Or something in between the two. Efficiency often correlates with effectiveness, although they are different measures. In terms of SAR response, greater effectiveness most often gives rise to greater efficiency, and similarly, greater efficiency gives rise to greater effectiveness.

4.3 (a) Efficiency - Man Overboard Response

The example of Man Overboard Response is a case where an effective immediate response will likely be a much more efficient system. After a person is seen to fall overboard, or believed to have fallen overboard, unless there is a quick resolution to the incident, generally the person will not survive, and in addition massive resources will be expended searching for the person (often their body). After an elderly man disappeared from the cruise ship MSC Divina near San Juan Puerto Rico, there was an extensive search, involving aircraft and a ship, the search was suspended after 35 hours.

Rescue crews searched for more than 35 hours, covering more than 2,000 square miles, according to the Coast Guard. Officials said that the search was conducted by an Air Station Borinquen MH-65 Dolphin helicopter aircrew, Air Station Cape Cod HH-60 Jayhawk helicopter aircrew, Air Station Clearwater C-130 Hercules fixed-wing aircrew and the crew of Coast Guard Cutter Donald Horsley, which is a 154-foot Fast Response Cutter home ported in San Juan [34].

In June 2016 a man was observed going overboard from the 29,000-ton ferry Spirit of Tasmania I, sailing in the Bass Strait between Tasmania and the Australian mainland. The ferry and her sister ship were joined by two other ships and several aircraft, after 12 hours a medical examiner deemed there was no chance of survival and the search was suspended [35].

There are about 20 Man Overboard cases per year from cruise ships and large ferries with a survival rate of about 20 percent. Of the 80 percent who perish, most likely die from cold shock in the first few minutes (other than those lucky enough to fall overboard into warm waters). For those who survive 'cold shock', rescue may be possible provided their position in known, and the rescue is timely [36].

4.3 (a) (i) Man Overboard Response - Unmanned Marine Vehicle Contribution

Response Capability - Can UMVs give rise to Greater effectiveness, thereby allowing greater efficiency.

There are fall detection systems available for ships, which can give timely notification of the event, and its position. Maintaining a clear indication of the victim's location and then early recovery is the best chance for a rescue, or, alternatively, the best chance for an early (body) recovery. Failing that, there will be a long search, requiring many resources. A UMV, or perhaps a 'marinized' UAV, could be launched once the Man Overboard alert is received (either automatically or by manual report). The craft would be launched to the recorded GPS location, use visual or IR cameras to look for the person in the water, and, presumably with the intervention of remote control from the ship, provide some flotation and encouragement, while clearly indicating the position for the ship's rescue boat. The UMV may be a very small craft, such as the U-Safe, a larger system such as the AGaPaS, or a rescue boat suitably converted for unmanned operation.

4.3 (b) Efficiency - Remote Area Response

Greater efficiency can give rise to greater effectiveness.

*"They (the crew) wanna know where the chopper was. They asked me, 'Where was the **** chopper,'... Watkins said his men spent hours in a life-raft before they were put aboard a cargo ship and then brought to Harbour Grace by another fishing boat... If it wasn't for those vessels, would there ever have been a chopper arrived in that area?" [37]*

As noted in section 4.2 (a) the fishing vessel Atlantic Charger sank in September 2015, and all the crew abandoned ship and fortunately were rescued by another fishing vessel.

Nunavut (Canadian Arctic) officials press for Arctic search and rescue base 'If we have a major incident up here, we are in a very bad position to be able to respond' [42].

A few months later, in February 2016, the heavily ice-reinforced shrimp trawler Saputi, while towing her nets at 3 knots, collided with a piece of ice estimated to be 5 M by 4 M and 0.5 M in height above the surface, and suffered hull damage leading to extensive flooding [Figures 8 & 9]. A Danish naval vessel departed Greenland to provide assistance, with an estimated time of arrival (ETA) of 18 hours. The Saputi's Master advised JRCC Halifax that the vessel was unable to keep up with the ingress of water using the pumps on board and requested additional pumps. A Hercules flew the 2,300 km from Nova Scotia and dropped four SAR pumps by parachute, these soon began to experience problems. The Zodiac from the Danish ship was damaged and unserviceable, the Saputi had 26 crew and 3 Danish naval personnel onboard. A SAR helicopter from Greenland was tasked with the evacuation of remaining personnel onboard, but it turned back due to icing. The Saputi's list increased to in excess of 45 degrees, with extensive free surface in the hold. The winds were from the northwest at 45 to 50 knots with a moderate swell from the southeast and poor visibility in snow. The Saputi eventually reached port in Greenland, escorted by the Danish naval vessel. All personnel survived, the Saputi was repaired and returned to service. *The outcome could have been very different* [38].

Figure 8: Saputi after striking ice (CAF) [42].

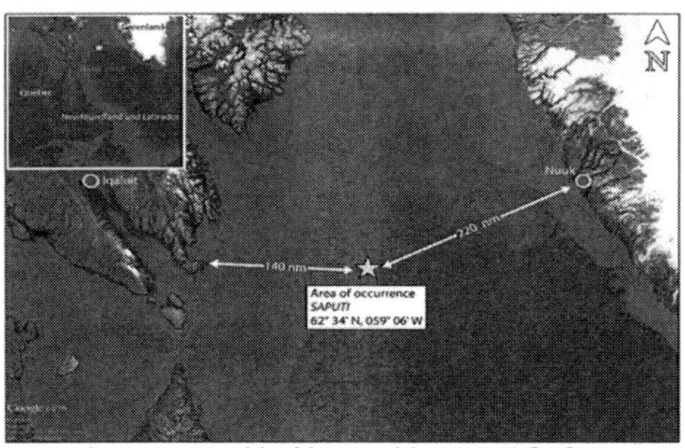

Figure 9: Saputi incident position (TSB) [38].

Canada has areas which are generally quite remote, have a small population base, relatively little shipping, and, fortunately, not many incidents. Other countries have similar issues. Under the SAR Convention some nation has responsibility for SAR coverage for even the most remote of sea areas.

As an example, the Halifax Search and Rescue Region (SRR) has more than 29,000 kilometres of coastline. It is an area of some 4.7 million square kilometres that is 80 percent water. This area includes:

- all four Atlantic provinces;
- the eastern half of Quebec;
- the southern half of Baffin Island in Nunavut; and
- the north-western quadrant of the Atlantic Ocean.

The Halifax SRR has the following primary CAF SAR squadrons:

- 103 Search and Rescue Squadron from 9 Wing Gander, Newfoundland. The squadron flies CH-149 Cormorant helicopters; and
- 413 Transport and Rescue Squadron from 14 Wing Greenwood, Nova Scotia. The squadron flies:
    - CC-130 Hercules tactical transport aircraft; and
    - CH-149 Cormorant helicopters.

Any other CAF or CCG unit may be used to back up the designated SAR units. They can be given the task of a SAR or SAR-like mission if they are: available, suitable, and capable. CAF assets are tasked to respond to approximately 400 SAR missions each year in eastern Canada. [39]

Consider the portion of the Halifax SRR Region north from Newfoundland - the Labrador Coast and the southern part of Baffin Island [Figure 10].

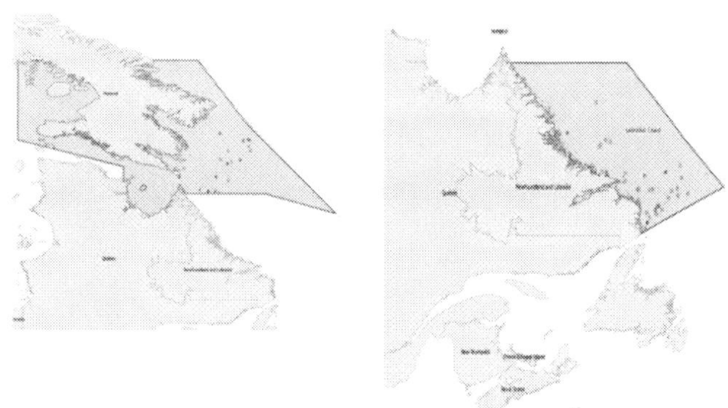

Figure 10: SAR incidents south-eastern Arctic/Labrador; 2009 – 2013 [43].

For the time period 2009 – 2013, the above charts indicate the location of various SAR incidents, Distress, Potential Distress, and Non-Distress. These are further noted in [Figure 11] below, for the Eastern Arctic and the Labrador Coast.

	Count	
Incident Severity Class	**Eastern Arctic**	**Labrador coast**
M1(Distress)	11	5
M2(Potential Distress)	7	7
M3(Non-Distress)	18	36
Total	36	48

Figure 11: SAR incidents south-eastern Arctic/Labrador; 2009 – 2013 [43].

As noted, there is a relatively limited number of incidents over a five-year period and spread over a large geographic area. However, there is a significant potential for incidents to occur. Small boats and fishing vessels are a significant part of the social fabric of these areas. There is also a considerable amount of shipping, raw materials, fuel and supplies, as well as a growing cruise industry. For instance, in neighbouring Greenland, there are 1,800 boats and vessels less than 120 tons (in a population base of less than 60,000 people), and nearly 40 cruise ships annually, some up to 4,000 passenger capacity (2011) [46, 47].

The Polar Code (in force 1 Jan 2017) has enhanced contingency planning guidance for passenger ships operating in areas remote from SAR facilities (Refer to MSC.1/Circ.1184). It states the following enhancements should be among those considered:

voyage pairing, i.e., mutual exchange of information that may be available to the SAR Authority or the vessel operator with reference to other passenger ships operating in the same area, so that, if two or more passenger ships are operating in the same general area at the same time, each can be used as a SAR facility in case of accident to another [40];

Transport Canada goes on to state:

Two groundings in the 2010 Arctic shipping season served as a reminder on the importance of contingency planning and risk assessment. As stated in the IMO Guidelines for Voyage Planning (A.893(21)), the detailed voyage plan should include, among other things, "contingency plans for alternative action to place the vessel in deep water or proceed to a port of refuge or safe anchorage in the event of any emergency necessitating abandonment of the plan, taking into account existing shore-based emergency response arrangements and equipment and the nature of the cargo and of the emergency itself." Access to emergency support services is very limited in Canadian Arctic waters. The shipowner may want to prearrange for emergency support prior to the voyage [41].

How do you provide sufficient SAR response capability for this large region, given the relatively low number of incidents, the harsh environment and the limited shore side infrastructure? In Canada most SAR resources are based much further south, aircraft and helicopters often have a thousand kilometers or more to reach the scene of the emergency, requiring time, allowing limited 'on scene' time and incurring significant cost in terms of operational expenditures and allocation of resources.
Canada has marine resources, in general Coast Guard or other ships are based in ports further south but may be patrolling in these waters and have a secondary role for SAR, effectively ships of opportunity. Canada is also establishing regional Coast Guard Auxiliary bases in the Arctic, these are smaller rescue craft, very useful in responding to small vessel emergencies, but perhaps of more limited capability regarding offshore and Mass Rescue cases.

4.3 (b) (i) Remote Area Response - Unmanned Marine Vehicle Contribution

Response Capability – Can UMVs give rise to greater efficiency, thereby allowing greater effectiveness?

Air Deployment - UMVs deployed from aircraft could assist with persons in the water, marshalling and maintaining communication with life rafts, standing by vessels in distress, and as a backup for SAR Techs parachuting into the sea. These UMVs would be small, and have limited capability, but could be equipped with a life raft, have communications links, and provide shelter and hope to rescuees. Control could be from the aircraft while overhead, transferring to another control station when the aircraft departs to refuel.

Increase the speed of conventional ships - Speed of response may be critical in emergency situations. Rescue organisations often try to address this with high-speed vessels. However, high speed in ships comes with a variety of design compromises, such as: high power requirements; light hull scantlings (structure); and, often, an impact on seaworthiness. Also, in heavy weather, the speed of these vessels may be restricted in order to reduce the accelerations and impacts experienced by the vessel and by the crew. The waters off Labrador and Baffin Island have a number of significant issues affecting the safe operation of high speed vessels; heavy sea states, navigational hazards such as pack ice, and more significantly, 'growlers' (small pieces of glacier ice, small enough to be difficult to spot in a sea, but large and hard enough to sink a ship).

As an alternative, a UMV could be carried by a conventional ship (with a speed of, say, 15 knots), such as a light icebreaker, buoy tender or emergency response tug, allowing a conventional ship to give a high-speed response to a SAR case, akin to that of a high-speed vessel. This UMV would likely be a rigid hull inflatable boat, perhaps nine metres in length, large enough to have the capability to travel some distance on its own at a reasonable speed, 25 to 35 knots (depending on sea state), small enough to be carried on a ship as part of the ship's equipment. It could be equipped with rudimentary accommodation for protection from the weather, communication equipment, and perhaps additional life rafts. This craft could act as an 'on scene' base for SAR Techs, could help respond should there be a Mass Rescue situation (such as a cruise liner in distress). Although prudence would be required for navigating in heavy seas at speed, should an unfortunate incident happen with the UMV, at least it would not be endangering ships' crews.

Stand-Alone Vessels - A Stand-Alone UMV could be prepositioned at sea or be based in a bay at a remote location. It could provide a significant survivor capacity and be a base of operations for SAR personnel, flown in to a nearby air strip, or perhaps SAR Techs parachuted into the sea or lowered from helicopters. If SAR helicopters are available, it could be a local base for helicopters to land survivors, allowing them to be more effective in a Mass Rescue situation (as was done in the case of the ferry 'Estonia').

A proposed outline description:

- Length 23 M, speed 20 to 25 knots, sufficient flotation to render unsinkable in case of substantial hull damage, hard chine hull design to allow simplified hull repairs, capable of operating in light pack ice, slush ice;
- Machinery package to give considerable redundancy; simplified maintenance including maintenance by replacement; arrangements to readily exchange old fuel;
- Limited live-aboard capacity for SAR personnel;
- Survivor capacity for UMV with associated life rafts – several hundred, rudimentary first aid facilities, including capability to treat hypothermic persons, helicopter winching area, small emergency boat capable of being remotely launched, controlled and retrieved;
- Designed to be readily boarded by persons in the water;
- Limited capacity for materials for pollution abatement;
- Full electronics package, including radio and satellite control/communications, search equipment including optical and infrared cameras, obstacle avoidance;
- Tethered drone/balloon/kite for raised antenna for search and communication;
- Fire monitor;
- Etc.

The stand-alone vessel could perhaps trail a passenger vessel as part of its contingency plan.

4.4 The Business Case – Unmanned Marine Vehicles for Search and Rescue

Is there a business case for the introduction of UMVs for SAR; i.e., from a Cost-Benefit perspective, do the Benefits outweigh the Cost? What are the barriers to their successful introduction and implementation, to maximize efficiency and effectiveness, while at the same time increasing the safety of SAR personnel? How can UMVs best be integrated into a comprehensive SAR system?

4.4 (a) Cost-Benefit Analysis Framework

A cost-benefit analysis is a process by which business decisions are analyzed. The costs associated with taking a specific action are calculated and subtracted from the benefits of a given business-related actions (including the savings from costs not incurred). Intangible benefits, in this case the saving of lives, will factor into the analysis.

The integration of UMVs into the current Maritime SAR systems will require better appreciation of UMV benefits including safety, effectiveness and efficiency. The maintenance, purchase, operating and acquisition costs will also need to be assessed. Therefore, the development of a business case that analyzes the cost-benefit factors will provide insight into the potential success of UMV use in Maritime SAR.

A concrete example would be the fitting of an improved Man Overboard (MOB) response system, incorporating a UMV, on a cruise liner.

Costs would include:

- Capital Cost – purchasing a UMV (or converting an existing rescue boat for unmanned operation), ship conversion such as launch and recovery apparatus, integrating into ship's electronics and navigation systems, regulatory approvals, and up-front training costs should also be capitalized.
- Operational Costs – maintenance, required periodic testing, ongoing training for ship's personnel
- Other costs

Quantifiable benefits could include:

- Direct benefit
  - Improved chance of saving lives
  - Positive public image of cruise line ("welfare and safety of passengers comes first").
- Costs avoided:
  - Saving of ship's time in case of person overboard (particularly if overboard detection system fitted).
  - Saving of potentially massive response costs by SAR authorities (although not directly a ship owner cost, government agencies may choose to mandate a MOB response system or threaten to charge ship owners for extensive SAR searches which may have been avoidable).
  - Savings from potential legal action in cases where available safety systems were not fitted.
  - Reduced risk to ship's crew to launch rescue boat in marginal or potentially dangerous conditions.

The improved chance to save a person's life may not be seen as an easily quantifiable benefit, but, hopefully, would be the most important benefit to any person of conscience.

This style of evaluation framework could be applied to the other possible applications of UMVs to SAR, with different inputs, benefits and costs. Once again, the main benefit may be lives saved, not easily quantifiable, but which is the raison d'être for all search and rescue organizations.

4.4 (b) Barriers to Introduction, Implementation and Integration into a Cohesive SAR Capability

Likely the greatest barrier to the introduction and integration of UMVs into the current Maritime SAR systems is a lack of understanding and appreciation of UMV benefits including safety, effectiveness and efficiency, and also of their limitations. Regulatory hurdles exist; however, IMO and many Flag Administrations are progressing in this regard. The successful integration of UMVs into a SAR system will require careful planning, and an intensive training program and operational knowledge, starting at the Management level and progressing down to the front-line operators. SAR technicians and crew members will need to know how to operate UMVs properly, otherwise the incorporation may be redundant and may in fact introduce additional hazards. Taking inspiration from the 'Unmanned Warrior' naval exercise, which has spurred the understanding, acceptance and integration of Unmanned Vehicles into naval applications, perhaps an 'Unmanned Rescuer' exercise would bring agencies together to do likewise and also capture the interest and imagination of the younger members of society.

5. THE WAY FORWARD

As long as people are involved with ships, and with the sea, Maritime incidents, endangering human life, will take place. Search and Rescue capabilities will have to evolve to meet these future requirements.

At the upcoming World Maritime Rescue Congress, Vancouver, June 2019, the theme will be "*Learning from the Past; Looking to the Future - Maritime Search and Rescue in a Dynamically Changing Operational Environment*". Unmanned Vehicles, air, ground and likely marine, will be part of this changing operational environment, to the benefit and Rescuers and Rescuees alike.

The technology is available, the naval experience is paving the way for the introduction of Unmanned Marine Vehicles. Costs in general are modest, and the case can be made that the benefits and the increase to safety will far outweigh the costs.

'THE APPLICATION OF UNMANNED MARINE VEHICLES TO MARITIME SEARCH AND RESCUE' should be given in depth consideration.

6. ACKNOWLEDGEMENTS

We thank Tara Sherrard, Industrial Engineering student, Dalhousie University, for assistance with background research, and formatting of this paper.
We also acknowledge the support given us by other members of faculty, by government agencies and by industry.

7. REFERENCES

[1] ROYAL NATIONAL LIFEBOAT INSTITUTION, 'Lifeboats', *1824: Our foundation*,
[2] SOLAS, 'SOLAS Chapter V – Contents', *Guidance on Chapter V – Safety of Navigation*, 2007.
[3] Almeida, J., Alves, J., Bertin, D., Dias, A., Ferreria, B., Ferreria, H., Fioravanti, S., Lobo, V., Martins, A., Matos, A., & Silvia, E., 'Unmanned Maritime Systems for Search and Rescue, Search and Rescue Robotics - From Theory to Practice', *InTech*, August 2017.
[4] KONGSBERG, 'The MBR Family', *Kongsberg Maritime*, 2007.
[5] MARITIME UK, 'MCA Code of Practice Version 1.0 Up to 24 Meters', *Society of Maritime industries*, November 2017.
[6] SHIPRIGHT DESIGN AND CONSTRUCTION, 'LR Code for Unmanned Marine Systems', *Lloyd's Register Group Limited*, February 2017.
[7] ROYAL NAVY, 'Unmanned Warrior', *Royal Navy Website*, 2018.
[8] DEFENCE SCIENCE INSTITUTE, 'Wizard of Aus. 2018 An Autonomous Warrior 2018 Trial – Applications Close 20th November 2017', *Defence Science Institute*, 2018.
[9] ASV GLOBAL, 'RIB Conversion', *ASV Unmanned Marine Systems*, 2018.
[10] ASV GLOBAL, 'ASV Global Delivers Autonomous System to The Canadian Hydrographic Service,' *ASV Unmanned Marine Systems*, 2017.
[11] MAPC, 'MAPC-20 Years of Unmanned Surface Vessel Work', *MAPC Maritime Applied Physics Corporation*, 2018.
[12] PELLERIN, C., 'Work: Robot Warship Demonstrates Advances in Autonomy, Human-Machine Collaboration', *Department of Defense United States of America*, April 2016.
[13] KONGSBERG, 'Autonomous Ship Project, Key Facts About YARA Birkeland', *Kongsberg Maritime*, 2018.
[14] KONGSBERG, 'Remotely-Operated Fireboat for Ports', *K-Magazine*, February 2018.
[15] ROLLS-ROYCE, 'Rolls-Royce Opens Autonomous Ship Research and Development Centre in Finland', *Rolls-Royce Media*, January 2018.
[16] ROLLS-ROYCE, 'Rolls-Royce Demonstrates World's First Remotely Operated Commercial Vessel', *Rolls-Royce Media*, June 2017.

[17] MAREX, 'Remote-Controlled Buoy Rescues Migrants', *The Maritime Executive,* May 2016.
[18] NORAS, 'U-SAFE', *Noras Performance,* 2017.
[19] CANADIAN COAST GUARD, 'Mid-Shore Patrol Vessel', *Government of Canada Website,* July 2015.
[20] MARITIME COURT OF BREMERHAVEN, 'Proceedings of 10th International Lifeboat Conference', *Maritime Court of Bremerhaven,* 1967.
[21] FLAHERTY, R., 'Investigations into Death of Coast Guard Volunteer Under Way', *The Irish Times,* September 2016.
[22] THE MARINE CASULTY INVESTIGATION BOARD. 'Interim Report of the Investigation into a Fatal Incident Involving an Irish Coast Guard Rib at Kilkee', *The Marine Casualty Investigation Board Report No. MCIB/266I/interim,* 2017.
[23] JOHNSTON, D., 'Death of Sergeant Janick Gilbert', *Governor General Website,* October 2011.
[24] ROYAL CANADIAN AIR FORCE, 'SAR Technician – A Cat – Epilogue – Flight Safety Investigation Report', *Government of Canada Website,* April 2014.
[25] COASTAL STATE FEDERAL REPUBLIC OF GERMANY, 'Fire on the Ro-Ro Passenger Vessel LISCO GLORIA', *Federal Bureau of Maritime Casualty Investigation Website,* February 2012.
[26] PENLEE LIFEBOAT STATION, 'The Penlee Lifeboat Disaster', *Penlee Lifeboat Station Official Website,* 2018.
[27] CANADIAN SHIPPING ACT, 'Life Saving Equipment Regulations', *Justice Laws Website,* March 2018.
[28] NEWTON, J., 'Lessons in Tragedy', *Royal Canadian Navy Website,* October 2016.
[29] UNITED STATES COAST GUARD, 'Mobile Offshore Drilling Unit (Modu) Ocean Ranger', *United States Coast Guard Website,* February 1982.
[30] THE JOINT ACCIDENT INVESTIGATION COMMISSION, 'Section 7.5 M/S Estonia The Final Report', *the Government of the Republic of Estonia Website,* December 1997.
[31] NEWTON, J., 'Inside the Harrowing Race Against Time to Rescue Nine Fishermen from the Middle of the Artic', *National Post,* September 2015.
[32] INTERNATIONAL MARITIME ORGANIZATION, 'Guide to Recovery Techniques', *International Chamber of Shipping,* November 2014.
[33] WIVB, '2 Missing Kayakers Make It Back Alive, 1 Found Dead', *WIVB News Website,* September 2017.
[34] SCICCHITANO, P., 'Coast guard Suspends Search for Cruise Ship Passenger', *Miami Patch Website,* January 2017.
[35] AUSTRALIAN MARITIME SAFETY AUTHORITY, 'Search Concluded for Man Overboard in Bass Strait', *Australian Government Website,* June 2016.
[36] KLEIN, R., 'Cruise and Ferry Passengers and Crew Overboard', *Cruise Junkie Website,* 2018.
[37] CBC NEWS, 'Owner of Atlantic Charger, Sunk in Frobisher Bay, Angry No Helicopter Responded', *CBC News Newfoundland & Labrador Website,* October 2015.
[38] THE TRANSPORTATION SAFETY BOARD OF CANADA, 'Marine Investigation Report M16C0016', *Government of Canada Website,* April 2017.
[39] NATIONAL DEFENCE AND THE CANADIAN ARMED FORCES, 'Search and Rescue (SAR) in Eastern Canada', *Government of Canada Website,* March 2018.
[40] INTERNATIONAL MARITIME ORGANIZATION, 'Enhanced Contingency Planning Guidance for Passenger Ships Operating in Areas Remote from SAR Facilities', *MSC.1/Circ.1184,* May 2016.
[41] TRANSPORT CANADA, '7A Voyage Planning for Vessels Intending to Navigate in Canada's Northern Waters', *Government of Canada Website,* 2018.
[42] ZEREHI, S., 'Nunavut Officials Press for Arctic Search and Rescue', *CBC News Website,* March 2016.
[43] SISAR: Search and Rescue Program Information Management System Database.
[44] ASV, 'Images', *ASV Global Website,* 2018.
[45] DARPA, 'Images', *Defense Advanced Research Projects Agency Website, 2018.*
[46] BERTHELSEN, K., 'Coastal Fisheries in Greenland', *KNAPK,* April 2014.
[47] ANTHONY, D., 'Cruise Ships in Greenland Waters', *Danish Maritime Authority,* June 2012.

8. AUTHOR'S BIOGRAPHY

John W. Dalziel, M.Sc., P.Eng., MRINA, studied Naval Architecture in the United States and Scotland, and Management Science in England. He has sailed through the Arctic with the Canadian Coast Guard, and with the fishing industry. Over the past half century, most of his experience has been in the supervision of ship construction, refits and repairs, and in safety inspection and survey on behalf of Flag State and Class. He has spoken internationally on maritime safety, most recently at the World Maritime Rescue Congress, Bremerhaven, Germany, 2015. As an Adjunct Professor in the Department of Industrial Engineering, Dalhousie University, Halifax, Canada, he is investigating 'The Application of Unmanned Marine Vehicles to Maritime Search and Rescue'.

Ronald Pelot, PhD., P.Eng., has been a Professor in the Department of Industrial Engineering at Dalhousie University since 1994. In 1997, he founded the Maritime Activity and Risk Investigation Network at Dalhousie. Since then his team has developed new software tools and analysis methods applied to maritime safety (accidents), coastal zone security, and marine spills. Research methods encompass spatial risk analysis, vessel traffic modelling, data processing and pattern analysis, location models for resource allocation, and safety analyses. Recent projects include developing an improved risk model of ship oil spills to determine zones of greatest risk depending on the degree of exposure, and the effects of extreme weather on fishing safety, and optimal resource allocation for Search & Rescue. Another study concerns shipping in the arctic, including a network model of feasible routes accounting for ice, land and bathymetry and a forecast of northern traffic in the future based on a wide range of drivers. Dr. Pelot has published over 50 journal articles and produced more than 100 technical reports.

Marine Propulsion Shafting configurations

M. Zeid, Caterpillar Propulsion, Sweden

SUMMARY

The marine propulsion drive train function is mainly to transfer the torque from Prime mover (Power source) to Power absorber (Propeller). The most important information to know at the beginning is the configuration of the power supplied to the system as well as location and type of propeller to be able to configure the shafting system in between. Considering the hull design itself from vessel to vessel also the vessel operating routes are different, this will influence the propulsion shafting configuration as well. The marine propulsion shafting configuration shall be well optimized in respect of suitable layout, minimum essential dimensions for components and possibility of installation. The paper is a conceptual paper where illustrating the most common shafting configurations for marine propulsion systems in three phases. The first phase "Layout" where considering different power and propulsion configurations for different vessels. The second phase "Dimensioning" where considering the main principles of different dimensions and sizes of the components for the same layout. The third phase "Installation" where studying the configuration from installation point of view, basically how to make it possible and easy to install in practical.

1. INTRODUCTION

The marine propulsion shafting having a high importance in the marine engineering. The reliability of the shafting system is essential, also the optimization to suit the hull design and the required function of the vessel with high mechanical efficiency is needed.

The marine classification societies has a minimum requirements for the main propulsion shafting system. However, the minimum requirement is to maintain the safe operation.

In order to optimize shafting system there is a traditional procedures where used to be followed. The more efforts into optimization process, the more efficient propulsion shafting in result.

The paper presenting the influencing parameters into the shafting configuration layout, dimensioning of the shafting with relative components and important consideration relative to installation.

2. SHAFTING CONFIGURATION LAYOUT INFLUENCING PARAMETERS

2.1 DIFFERENT VESSELS

2.1(a) Hull Shape

There are many influences of the hull design into the selection of the propulsion type and relative number of main propellers.

One of the common influences is the submersed hull shape. The submersed hull shape is one of the major contributor of the propulsion type. The number of propulsors is influenced by hull type where it can be monohull, catamaran or trimaran,.. where likely the number of the main propellers (1, 2, 3 respectively) equipped to a specific vessel is in line with the hull type.

One other common influence is the available draft of the designed hull is also contributing the number of propulsors as well as the size. In such hull designs where the draft is limited the number of main propellers mostly more than one with relative less propeller diameter in order to reach the desired output for the one propeller equivalent propulsion.

2.1(b) Vessel Function

The influence of the vessel function into propulsion type selection. The speedy vessel require a high hydrodynamic optimization for the hull specially on the aft part as well as the propeller, in such kind of vessel function is kind of unlikely to have an external propulsion type (example, vertical axis propeller, azimuth thruster,..)

This can be the opposite for the vessels where subjected basically for delivered bollard pull and operating in an operational risk area (big object in the water expected to stuck into propeller regularly). In such case external propulsion system where easy to replace can be more preferable.

2.1(c) Input Power To The Propulsion System (Prime Mover)

Marine propulsion power configuration is a very wide matrix for today configurations. This as the integration of electric motors, generators and internal combustion engines are widely used for optimizing the power consumption toward more green solutions in marine industry by the meaning of hybrid systems and similar. In that sense the input power to the main propulsion has been represented by "Prime Mover" where it can reflect electric motor or internal combustion engine driven or equivalent.

When the type of the propulsion is identified, the input power source (main engine or electric motor) can be selected according to the power range required as well as operating profile [1].

2.1(d) Marine Reduction Gear

The Marine prime mover basic RPM (revolution per minute) categories considering similar power range:
High speed, where the prime mover revolution per minute likely 1600 RPM and above.
Middle speed, where the prime mover revolution per minute a range around 600 – 1000 RPM.
Low speed, Less than 200 RPM.

With reference to marine propeller theory [2] and in order to have optimum propeller design, a reduction gear is likely needed except for low RPM rated prime movers.

2.1(e) Prime Mover And Reduction Gear Layout

The longitudinal and vertical location of prime mover and reduction gear reference to propeller has a high influence into the shafting layout.
The shaft dimensioning for marine application is governed by IACS (M68) formula [3].

With reference to IACS formula, the less torque, the less shaft diameter is needed. Respectively, the part of shafting between prime mover and reduction gear will be less dimensions in compare to the part of the shafting between the reduction gear and propeller. This is one of the major influences of the reduction gear longitudinal position into the shafting configuration.

The thrust from propeller is developed by normal operation of the propeller. This thrust need to be supported by a bearing. The most common location of the propeller thrust support bearing is inside the reduction gear.

As per natural physics, it can't be any high axial flexibility between propeller and propeller thrust support bearing (Thrust Bearing). Respectively, all high flexible elements where can induce high axial flexibility are located between the prime mover and reduction gear (Figure 1).

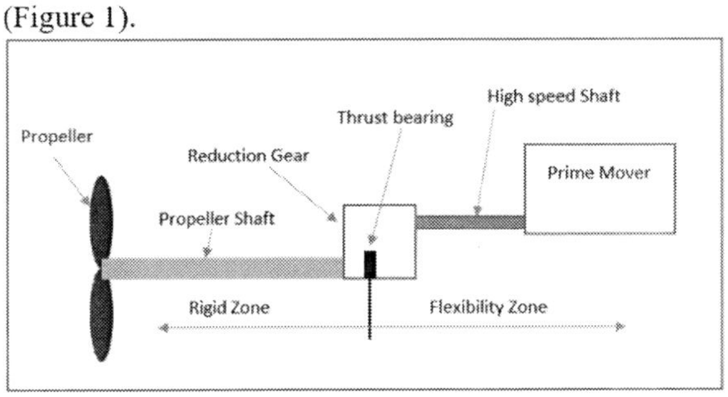

Figure 1: Schematic layout for rigid / flexibility zones.

2.1(f) Flexibility Degree

There are different degrees of flexibility can be applied to the shafting configuration. The vertical offset of the prime mover toward the gearbox is influencing the type of flexibility required for shafting configuration.

Torsional flexibility:
This presented by torsional flexible coupling between the prime mover and the reduction gear. This type of flexibility is suitable for configurations where the prime mover located longitudinally close to reduction gear, also applicable when no vertical offsets between prime mover and reduction gear.
The torsional coupling provides the essential torsional flexibility and damping to the system in respect of torsional motion.

Torsional flexibility and moderate radial / axial flexibility:
This presented by torsional flexible coupling and torsional / shear stiff elements integrated to the shafting line between the prime mover and the reduction gear. This type of flexibility is suitable for configurations where the prime mover located longitudinally far from the reduction gear. Also, for configurations where prime mover or (and) reduction gear installed in flexible mounts to the relative foundation where can be suitable in this case for short longitudinal distance between prime mover and reduction gear as well.
This type of flexibility is applicable when no or slight vertical offsets between prime mover and reduction gear is presented.

In addition to the benefits of the torsional coupling, each flexible element (torsional / shear stiff element) provide limited value of bending and axial flexibility. Each two of these elements provide radial flexibility into the shafting. In that way the reflection of hull girder deflection into shafting radial support bearings are decreased significantly.

Torsional flexibility and high radial / axial flexibility:
This presented by torsional flexible coupling and cardan shafts integrated to the shafting line between the prime mover and the reduction gear. This type of flexibility is suitable for configurations where the prime mover located longitudinally far from the reduction gear. Also, for configurations where prime mover or (and) reduction gear installed in flexible mounts to the relative foundation where may be suitable in this case for short longitudinal distance between prime mover and reduction gear as well.

This type of flexibility is applicable when high vertical offsets between prime mover and reduction gear (Figure 2).

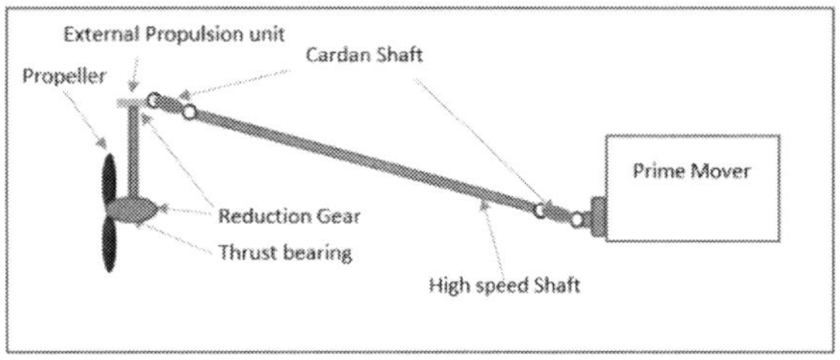

Figure 2: Schematic layout for high radial / axial flexibility.

In addition to the benefits of the torsional coupling, each cardan shaft provide high value of bending and axial flexibility. In that way the reflection of hull girder deflection into shafting radial support bearings are widely decreased. However, the cardan shaft induce a relative high value of forces [7] where subjected to second shaft order. In that sense the integration of the cardan shaft has to be well studied.

2.2 DIFFERENT PROPULSION TYPE

2.2(a) Conventional Propulsion

The conventional propulsion solution (Controllable Pitch or Fixed Pitch propellers) is a solution where optimized for high level performance where used to provide a high efficiency for propulsion systems. In this case all the components of the system has been optimized for the specific vessel with the specific function at rated power.
The system consist of main propeller (Controllable / Fixed pitch), stern tube with relative bearings and seals. This in addition to the shaft extension inboard where coupled to the reduction gear and then to Prime mover. Additional power management components may be integrated to the system in parallel (additional branches at reduction gear) or in series (in main propulsion shafting lines).
This type of propulsion is more suitable for high performance vessels where the high efficiency with relative low fuel consumption is required.

2.2(b) External Propulsion

The external propulsion solution (Azimuth thrusters, Azipods as an example) is a solution where optimized for compact design where used to provide a good efficiency levels for propulsion systems. This as the propulsion unit contains the main propeller, reduction gears and the azimuth function (turning 360 degrees) provides a high maneuvering performance as well as no need of rudder in behind propeller.

The expression of External propulsion is the author expression for simplification of categorize different type of propulsion.

The propulsion unit used to contain most of main components together where optimized for best performance as a system. Additional power management components may be integrated to the system in parallel (additional branches inboard at reduction gear) or in series (in main propulsion shafting lines).
This type of propulsion is more suitable for vessels where the high maneuvering level required, specifically when the vessel rated as "Dynamic Positioning" system where the high maneuvering level of the propulsion unit lead to not be in need to add transverse tunnel thrusters at the aft part of the vessel. This in addition to the vessels where subjected to daily routs operations (Ferries, Tug boats,...) where the down time is limited.

This as the entire propulsion unit can be refitted easily to hull with a new unit and repairing defected unit out of the vessel and lead to getting back to operation in significantly less time compare to conventional propulsion as the repair job for this type can be carried out in land.

2.2(c) Combined Propulsion Systems

The Twin Fin is a patented solution that where presented to the marine market for the first time in 2012 by Caterpillar Propulsion solution. The solution is a combination between conventional and external propulsion.

It combines the benefits of conventional propulsion that are simple, robust and where most components are accessible from inside with the Azimuth / Azipod where you are free to positioning the propeller most favorable and have most components outside the main hull.

As the Azimuth / Azipod the Fin and propulsion system from propeller to motor, can be prefabricated in controlled environment and installed using minimum time in dock (Figure 3).

The optimization principle for the Twin Fin solution is to install very simple and compact conventional propulsion package in a well-tuned hydrodynamics optimized fin giving you a highly efficient long life propulsion with protected propeller (Figure 4), no bearings outside and as side effect you will gain extra deadweight.

Figure 3: Installation of Twin Fin.

Figure 4: Twin Fin layout model.

2.2(d) Pre Designed Propulsion Systems [4]

For certain type of vessels where having the same function and the major change in the principle dimensions is limited (Tug boats, Service boats…). It used to be a pre design performed by propulsion makers where the system is well optimized mechanically and propeller design is designed to suit the specific vessel.
The major benefits of such systems is the ready configurations, high mechanical efficiency with proven performance and ready installation manual.

2.3 SHAFTING LAYOUT SKETCHING

At this point the main principles of propulsion shafting configuration are available. Respectively a schematic layout of the propulsion shafting line can be established.

3. SHAFTING CONFIGURATION DIMENSIONING INFLUENCING PARAMETERS

The nominal torque transmitted through the shafting is the governing parameter for dimensioning the shafting components. However, for the same shafting layout where subjected to certain value of nominal torque, the shafting dimensions and components size can differ significantly. The most common influencing points are discussed below.

3.1 SHAFTING SYSTEM DYNAMICS

The main propulsion system dynamic responses in different motions shall be simulated in order have a clear view of the system dynamic behavior.

3.1(a) Torsional Analysis

The propulsion shafting system shall be examined toward torsional vibration where to find if any excessive torsional response. The major aim of this analysis is to dimension the required characteristics of the torsional vibration coupling and torsional damper (if needed).

Once the minimum required characteristics for the coupling identified, the coupling size maybe chosen differently according to the expected risk value to dominate.

The entire system components shall be verified in according to the results of the torsional analysis, each component in according to its own criteria.

The results of torsional analysis influenced components with interface in lateral or (and) axial motion (Propeller, gears, cardan shafts…), shall be considered in the lateral / axial analysis modeling. This to have the analysis results more reflective to real operation.

The common marine torsional analysis tools are considering simplified mass elastic data concept (Figure 5) for modelling. The frequency domain and time domain analysis are also available.

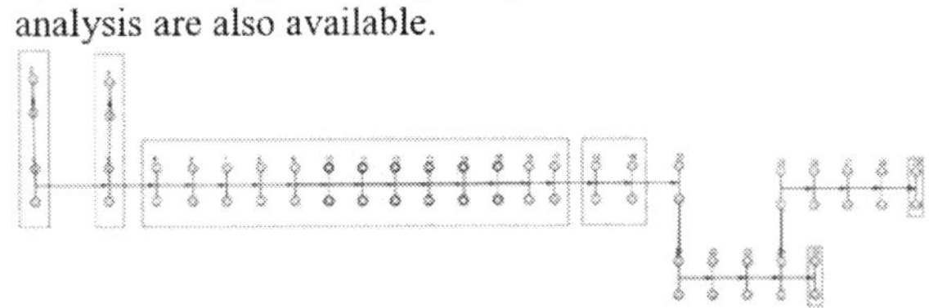

Figure 5: Torsional Vibration analysis model.

3.1(b) Lateral / Axial Analysis

The Shafting system shall be subjected to lateral and axial analysis. This to study the shafting behavior in lateral and axial.

The most of available tools are considering finite element modal analysis for the system where make it easy to carry out the lateral and axial in one model (Figure 6).

Figure 5: Lateral / Axial Vibration analysis model.

The bearing loads and relative natural frequencies are the major outcome of the analysis tool.

The modeling shall consider the relative hydrodynamic forces and torsional lateral (and axial) interfaced forces to the model in different operating conditions in order to examine the system in the expected working domain.

When the shafting system consist of more than two radial bearing in series, the shafting system is considered statically indeterminate system. In order to balance the loads between bearings, changing the bearing offsets is applicable.

The bearing size selection is depending on bearing loads obtained from analysis, where the stationary load (static) and dynamic load with relative frequency (harmonic) need to be considered during bearings calculations. Selecting the suitable bearing size (for same shaft diameter) is applicable for roller bearings. For slide surface bearings, a customized bearing design [6] can be applicable if needed.

The shaft deflection as one of the output from this analysis, helps to identify the type of seals where suitable to use for the application.

The relative high speed propulsion shafting (for a propeller less than 1.5m) is more sensitive toward lateral vibration [5] and relative low speed propulsion shafting (for propellers bigger than 1.5m) is more sensitive toward bearing offsets.

The required thrust bearing stiffness is one of the output where obtained from this analysis as well. It is highly important in case of flexible mounted reduction gear.

3.1(c) Higher Level Simulation

Coupling between torsional and axial vibration analysis or (and) flex body analysis maybe needed in case of new conceptual design.

3.2 MARINE CLASSIFICATION

The marine classification requirements can influence the shafting configuration significantly. Considering same shafting layout for two vessels, one vessel with no ice class and the other vessel with high ice class. The shafting diameter and shafting components for the vessel with high ice class will be bigger than the no ice class vessel shafting diameter and components.

3.3 OPERATING ENVIRONMENT

The operating environment can influence the shafting configuration. For example, operating in a shallow water in a river can indicates a risk of getting wooden pieces (or similar) stuck between propeller and propeller nozzle. Respectively, a suitable clutch need to be integrated to the shafting system to avoid excessive damage.

4. SHAFTING CONFIGURATION INSTALLATION CONSIDERATIONS

4.1 SHIPYARDS CAPABILITY OF INSTALLATION

Checking the shipyard capability for handling the proposed shafting components is essential. It can happen that some of the components not suitable for installation by shipyard.

4.2 SENSITIVE INSTALLATION

As mentioned in Lateral Analysis, the bearing offsets maybe having offsets in order to have suitable load for each bearing. During installation, the system installation shall maintained the specified offsets.

The GAP and SAG method is the most common way in marine application for applying offsets to the shaft bearings or reduction gear. By adjusting the forward flange of each flange coupling (while flanges open) to a specified GAP and SAG values toward aft flange. The bearings will be located at desired offsets respectively.

Checking the bearing loads by the meaning of jack load test is the most common way for confirming the correct installation for the shafting components inside the vessel.

When the shafting configuration is sensitive to installation, the marine classification society requires "Alignment Report" where basically the lateral shaft analysis report in addition to pre described the GAP-SAG and Jack load test values as well as recommended procedures to carry out the alignment.

5. CONCLUSIONS AND RECOMMENDATIONS

It is important to put essential efforts into configuring the propulsion shafting system.

Considering the configuration is well mechanically optimized scaling from other similar configuration can result in major failures in the propulsion shafting components. This as each system behave differently due to the wide range of influencing parameters.

In case of pre designing a specific shafting configuration, it is essential to consider a range of change and examine the system along this range. This in order to build a trusted domain.

When the system defined as "Sensitive to installation", it is essential to follow strictly the installation guidance provided by the maker.

6. ACKNOWLEDGEMENTS

I would like to thank my family for all support I get to do this work. I'm also grateful to Caterpillar Propulsion where supported with all needed resources to finalize this job.

7. REFERENCES

1. 'Caterpillar Marine Engine Selection Guide', 2013.
2. Carlton, J. S., 'Marine propellers and propulsion', *Elsevier Ltd.*, 2007.
3. 'IACS Requirement 2005', M68 Rev.1, 2014.
4. 'Caterpillar Marine Engines Application and Installation Guide', 2005.
5. Veritec, 'Vibration Control In Ships', *A.S. VERITEC*, 1985
6. M. Zeid, 'Sailing conditions effects into Propeller Bearing for big vessels', *Caterpillar Propulsion internal researches*, 2013

7. Gewes, 'Product catalogue', Gewes website, Internet.

9. AUTHORS BIOGRAPHY

Mohamed Zeid I'm currently holding the position of Rotordynamics specialist at Caterpillar Propulsion. I'm responsible for Rotordynamics group activities within core competence team since 2006. As originally a Naval Architect and Marine Engineer, my previous experience includes intact / damage stability, structure analysis and strength calculations, machinery configuration and arrangements, pipelines design, etc.

UNDERSTANDING ENGINE POWER CHARACTERISTICS AND THEIR EFFECTS ON BOAT PERFORMANCE AND ENGINE RELIABILITY IN LOW HOUR, HIGH LOAD, RNLI SAR LIFEBOAT APPLICATIONS

S Johns, Royal National Lifeboat Institution, UK

SUMMARY

This paper discusses diesel engine power characteristics, with reference to specific case studies, focusing on details such as engine power and torque curves, and how small changes to these and driveline components such as propellers can have a big impact on engine performance and reliability. Methods and approaches to try and control these changes and solutions developed to achieve the required performance and reliability through modifications and engine selection are discussed.

NOMENCLATURE

ALB	All weather Life Boat
ECM	Electronic Control Module
High idle	Maximum engine rpm without load
hp / bhp	Horse Power / Brake Horse Power
Knts	Knots
kW	Kilowatt
L	Litre
MEUI	Mechanical Electronic Unit Injection
Mk	Mark / revision
Rated rpm	Speed at which the engine achieves rated power
Rated Power	Maximum power produced at rated rpm
RNLI	Royal National Lifeboat Institution
rpm	Revolutions Per Minute
SLEP	Severn Life Extension Project

1. INTRODUCTION

RNLI lifeboat designers have long been stuck selecting a high-performance pleasure rated engine for maximum power to weight ratio, whilst trying to achieve continuous duty rating reliability.

Not understanding, how an engine achieves its rated power, power tolerances and effects of adjustments to performance parameters, can have significant implications for both boat performance and engine reliability. In addition, boat modifications can have similar consequences for engine performance. This becomes more pertinent where long fleet build periods and long service lives, allow more opportunity for change to occur. Conversely, with advances in technology and manufacturers taking different approaches to power control, these changes can also be used to your advantage.

2. POWERING RNLI ALB LIFEBOATS

2.1 RNLI SAR HIGH POWER DEMAND

A question that is often asked is why do RNLI lifeboats have so much power for 25 knts, when compared to other boats?

The main reasons are the RNLIs unique operational requirements, such as the need for the ALBs to be able to take the ground, which in turn lead to significant performance constraints.

Features such as heavily tunnelled propellers and large bilge keels restricting water flow to the propellers and the subsequent restriction on propeller diameter, result in the need for a high engine power density to achieve 25 knts, nominally requiring up to 30% more power when compared to similar sized boats without these design features.

This high power density demand and high power to weight ratio requirement forces engine selection into the pleasure rating category.

2.2 RNLI OPERATION

Although our operating hours are low, typically 200 - 250 hrs a year, the portion of this time spent at high loads can be comparatively high spending longer at rated power than the manufacturer would normally recommend. With engine manufacturers guidelines on usage for a pleasure rating being typically 8 – 20% of time at rated rpm, we can see up to 50%.

A typical RNLI load profile being split between low rpm and rated rpm with limited operation in the range in between as shown in figure 1.

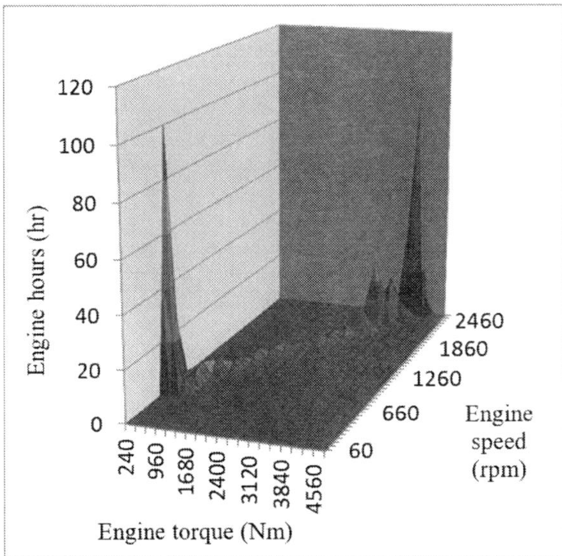

Figure 1: [Download of an RNLI ALB engine load profile]

This then adds in additional complexities such as bespoke maintenance regimes with increased component exchange rates and reduced overhaul periods, and reduced time between major overhauls. All of which lead to significantly increased maintenance costs and boat down time.

The RNLI rarely install an off the shelf engine, with some form of hardware and / or software modification being required, particularly regarding capsize capability.

Being the boat owner, operator and maintainer, and with an expected service life of 25 years, performance and reliability become increasingly important.

With a class fleet size of around 40 boats, production tolerances can creep in and engine loads can go up, with power demand between boats varying by as much as 5%, making the power reserve at the design stage ever more important.

The issue of trying to keep a fleet of boats the same for advantages such as commonality of spare parts, operating procedures and training can also lead into issues such as 'last time buys' towards the end of fleet production as the engines are no longer available, with the added complication of storing a large quantity of engines and how to maintain them before installation.

With additional components such as propellers being standard across the fleet, and not optimised for each boat, also adding to the variances in power demand and performance.

2.3 ENGINE DEVELOPMENT

With the process of design concept through to production of a new class of lifeboat taking up to 5 years or more, by the time the first boat enters production the engine initially selected or looked at is more than likely obsolete or has been superseded. Even where the same model may appear to be current, it can be a completely different engine just with the same model number.

Typically derived from land-based applications before being converted for marine use, marine engines nominally only have a 5-year production life before being replaced with a newer model.

Choosing a new engine to maximise availability is the preferred approach but does bring with it the risk of being a 'test bench' for the new engines. You don't want to be regularly visited by the manufacturer for continual trials and testing trying to correct issues as the engine is released to market.

Throughout their production life marine engines are constantly being revised and developed. Even when buying the 'same' engine, you can almost guarantee there will be differences somewhere between each unit, either mechanically or within the software.

Key items are often superseded and upgraded and this can be anything from performance enhancements through to changes in components to rationalise parts used on the production line.

2.4 ENGINE RATINGS

Ratings for high speed marine diesel engines typically range from heavy duty continuous use through to a pleasure rating which have time limitations on operation at rated power. The power output of a pleasure rating for the same engine often being more than twice that of the heavy duty rating.

To achieve these different power ratings, it can be as little as just the fuel limits that are adjusted, or in addition, sub components such as pistons may be changed along with increased cooling capacity to cope with the increased power, the core block and running gear often remaining the same.

You also need to pay attention to the small print. If you end up with an application that does require every last hp to achieve performance, then the power test limit tolerances at engine production, typically +/- 3%, could potentially result in a performance hindering power loss.

Obviously, this situation should be avoided where possible at the engine selection stage.

2.5 OVERRUN CURVE

Whilst engine performance data (horse power and torque figures) for each rating is normally readily available, what is not quite so easy to come by and often overlooked is the data for the overrun curve and how the power is reduced between rated rpm and high idle.

Where used, an overrun curve gives a controlled reduction in power between rated rpm and high idle. This in effect allows a sea margin or for hull fouling, giving the engine a reserve of power to maintain rpm when load increases.

Originally derived from mechanically governed engines its characteristics are often designed into the performance curve of a modern electronic engine. Its function to give stable power control above rated rpm, where varying load demand would otherwise result in uncontrolled oscillating rpm if the fuel were to be simply cut off. However, with modern engines, with their millisecond governor response times and electronic injectors, this is no longer a necessity.

The engine manufacturer will advise to operate the engine in the 'preferred rpm range' where possible, the rpm range typically being 1 – 3% above rated rpm (Figure 2).

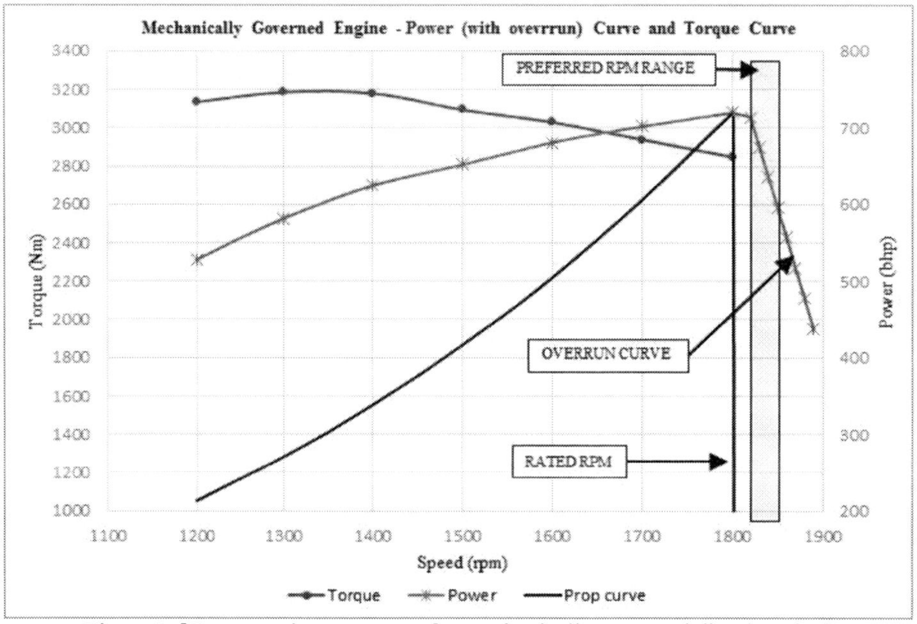

Figure 2: [Torque and power curve for mechanically governed diesel engine]

2.6 TORQUE CURVE PROFILE

A typical torque curve rises to its peak value in the middle of the rpm range reducing back down towards rated rpm (or capped on fully electronic engines). This means that as engine load increases, and rpm reduces below rated rpm, the torque increases helping resist any further rpm reduction (Figures 2 & 10).

3. THE ISSUE WITH HIGH POWER DENSITY

It would be fair to accept the assumption that the core structure of an engine is a solid and rigid assembly that doesn't move. However, this is not the case and they can flex more than would maybe first anticipated. In order to transmit the power and torque produced to the vessel, and retain the pressures and temperatures required, the engine has to resist these forces in equal and opposite measure.

When a manufacturer starts talking to you about revised blocks, ladder plates or stiffened sumps, it is likely they are struggling to contain these forces.

If you ever have the opportunity to view an engine under load with a stroboscope, you may be surprised to see how much parts actually flex and move.

This is why it is wise to be cautious with engine models that share the same core components but have ever increasing power outputs as they are updated to remain competitive.

These higher powers and increased loads, increase the temperatures and pressures and resultant stresses on an engine and this is exacerbated when an engine is overloaded or running in 'Lug'. Where high output pleasure ratings are concerned, these may be at the limits of the core engine

Ultimately there is only so much life in an engine either from long hours at low power or low hours at high power.

4. CASE STUDY 1 - SEVERN ALB MK1

LOA:	17M
Displacement (t):	42
No. of engines:	2
Propulsion:	4/5 Blade propeller
Engine configuration:	V12
Capacity (L):	27
Rated power (hp):	1250
Rated rpm:	2300
Cycle:	4 Stroke diesel

4.1 BUILD HISTORY

The Severn class entered service in 1995 and continued in production through to 2004 to complete a fleet of 45 boats. The issues around the performance problems of the class were quite complex and took some effort to resolve.

It is worth mentioning at this stage that the weight of the class had increased from the preliminary design weight of 36 tons to 42 tons in final production and service, due to various design changes and modifications.

Throughout the boats development stage the engine power had also gone up, increasing from 1050hp up to the 1250hp rating that was finally fitted to production boats.

It should also be noted that although technically classed as an electronic engine, with an ECM governor for power management, it was actually a mechanically fuelled engine with a mechanical scroll fuel pump and mechanical injectors. This gave significantly different power characteristics from the 'Common Rail' or 'Unit injection' engines we know today.

4.2 PERFORMANCE ISSUES

At new build commissioning the first 20 or so boats met acceptance criteria and were coming out with the engines reaching rated rpm and entering into the 'preferred rpm range' on the overrun curve at around 2330 rpm. But, as production moved into the second half of the fleet the engine rpm started dropping off, with rated rpm only just being achieved at full speed.

Then we started to experience performance issues across the fleet, particularly once the later boats had been in service for 6 months or more.

The engines were now failing to achieve rated rpm, often running at around 2260 - 2280 rpm at full throttle and would often hang at around 2060 rpm, being stuck until a suitable following wave would lift the boat and allow them to accelerate again.

Whilst the symptoms were similar between boats suggesting a common problem, some boats appeared to be less affected or completely unaffected. At the same time, we started to see an increasing rate of engine failures.

4.3 INVESTIGATION

In the early part of the investigation it transpired that a propeller design change halfway through the build program had been introduced, primarily to resolve severe root erosion due to cavitation.

Initially this did not give cause for concern, the change being described as only a hub modification (The propeller hub being extended and its diameter increased), the pitch and diameter staying the same (Figure 3).

Figure 3: [4 blade propeller with modified hub showing $CaCO_3$ deposits]

Later in the investigation it transpired that although the engines had been achieving 2330 rpm on the earlier boats, the boats had been struggling to achieve their maximum speed of 25knts.
In an attempt to restore boat speed, it had been communicated to the propeller manufacturer that there was a "spare 30 rpm" above rated rpm that could be used to absorb more power and this was factored into the propeller design change.

By modifying the propellers to use up this rpm gap the engines were now working at full load to achieve rated rpm. The power margin had been completely removed and with any additional loading, such as hull fouling the engines were placed in an overload condition.

This could be caused by as little as just a thin layer of Calcium Carbonate $CaCO_3$ that would typically build up on the propeller blades (figure 3).

At this point, an interim solution for these particular boats being to either acid clean the propellers to remove the scale at an early haul out, which would often return some performance or fit the earlier propeller design to reduce some load and accept the speed loss.

However even the earlier propeller design could not fully resolve engine performance loss even on earlier boats once the boat weight had increased and the propeller and hull were fouled.

4.4 THE ENGINE TORQUE CURVE

Re-visiting the torque curve for the engine fitted also started to bring the pieces together.

Rather than rising, the torque curve dropped away from rated rpm, before beginning to rise again below 2100 rpm coinciding with the point at which some boats would hang (Figure 4).

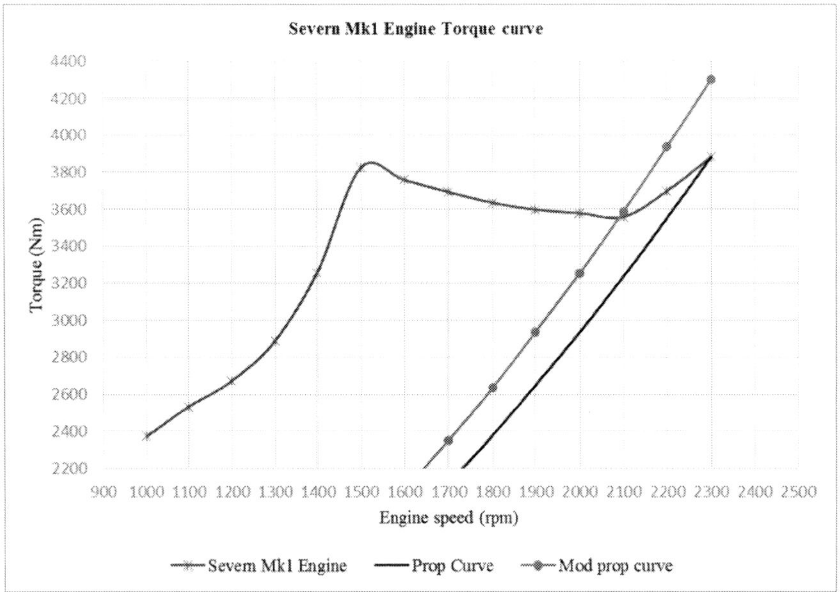

Figure 4: [Torque curve of engine fitted to Severn Mk1 ALB highlighting intersection at dip with modified propeller demand curve]

This meant the engines were heavily overloaded, running in a 'lug' situation with the fuel rate on its limit for that particular rpm.

This was also backed up by the sea trial measured fuel burn rate figures only just coming back into the performance band limits at this point (Figure 5).

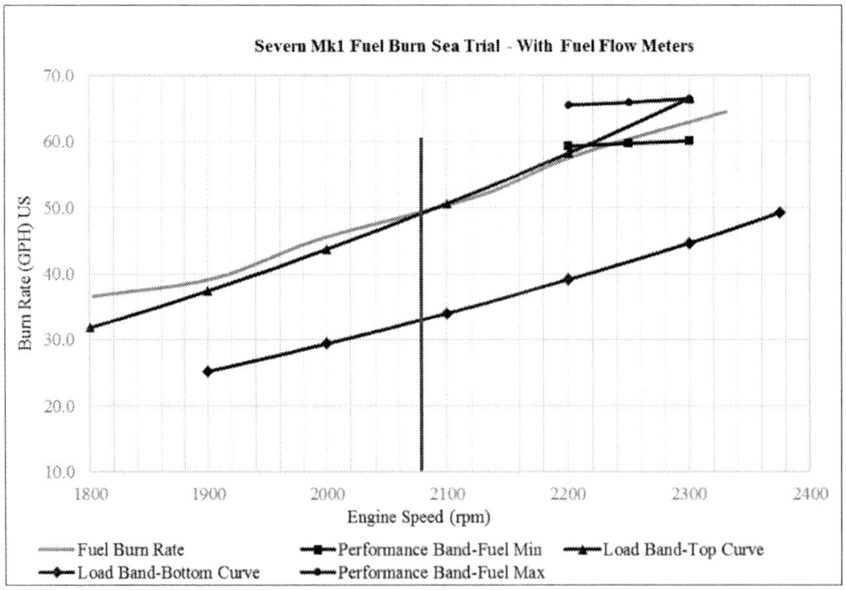

Figure 5: [Fuel burn rate from Severn Mk1 ALB sea trial with modified boss propellers]

4.5 ADDITIONAL ISSUES

By the end of the investigation other issues had also been uncovered. With the engine having an electronic governor it was possible to change the power output on the engine directly with a laptop, by adjusting the fuel rate limit.

Service engineers, when attending a boat with performance issues, were in some cases, simply adjusting the fuel limit until the required performance was achieved, without fully understanding the consequences. This was compounded by some engines having different power rating options available (1-4 or 1-5), with varying fuel limit parameters, as the software was developed through its production life. The relevance of these ratings being misunderstood and deemed more important than the actual fuel rate limit setting itself.

When some engines where placed back on the dyno following major overhaul, they were found to be producing as much as 1350hp, 8% more than the rated power, due to uncontrolled adjustments.

With the tolerance for the engine power output being +/-3%, this could result in a total reduction from rated power by as much as 75hp per boat, if both engines were at the lower limit. This had the potential for a 150hp difference between boats with the installed power ranging from 2425 – 2575hp, just on the manufacturers tolerance alone. Again, as some engines passed through the dyno they were found to be on the lower limits of power output.

Although it would be unlikely to arrange for the manufacturer to tighten this power tolerance at production, it is possible with modern facilities to achieve tighter power tolerances with an engine on a dyno, particularly in our case where they are often re-tested.

This variance in power output, both from the manufacturer and local adjustment explained why some boats suffered more than others with performance issues and some appeared to have no problem at all.

4.6 THE SOLUTION

The final solution to re-gain performance across the fleet and reduce engine failures required several steps.

- Following extensive trials, a new 5 blade propeller (Figure 6) was introduced across the fleet. The revised design, with an increase in diameter over the original 4 blade, but with a reduction in pitch, reduced engine loading to around 98% at full speed. More importantly the boats were able to maintain engine performance between the prescribed 6 monthly haul outs.

Figure 6: [New 5 blade propeller for Severn Mk 1 ALB]

- The engine power output tolerance, in agreement with the engine dealer, was tightened to +3% only. When ever an engine was re-tested on a dyno following repair or overhaul, its power output was set to 1287hp.

- Both RNLI and the engine dealer service engineers were stopped from making any rating adjustments on the engines without express authorisation. Any in service engine that required performance investigation or adjustment, being attended to by specific personnel with dyno calibrated fuel flow meters to measure power output.

- Introduction of the 20s plenty rule. Crew were educated to the fact that even though it was classed as a 25 Knt boat, there were implications for engine reliability if full speed was used excessively, with 20 knts being the advised maximum speed unless on an emergency call out.

5. CARBON PROPELLERS

During development of the revised 5 blade bronze propeller (Figure 6), some considerable time and effort was spent procuring and trialling a pair of carbon fibre propellers as another potential method of reducing the load on the engines.

The theory and main advantage being their ability to flex under load. During hard acceleration the propeller would effectively de-pitch reducing power demand allowing the shaft rpm to increase, pitch returning and power demand increasing as rpm stabilised.

There were other benefits as well, such as a 60% weight reduction over all bronze and the ability to replace single blades if they became damaged.

Although initial trials were very positive with a significant improvement in acceleration to full speed and a reduction in engine loading, they could not withstand the hard-astern manoeuvres, rapidly fracturing and shedding the blade tips. Although not progressed further at the time, this is definitely a technology worth keeping an eye on for the future.

6. CASE STUDY 2 – TAMAR ALB

LOA:	16M
Displacement (t):	32
No. of Engines:	2
Propulsion:	5 Blade propeller
Engine configuration:	I6
Capacity (L):	18
Rated Power (hp):	1000
Rated rpm:	2300
Cycle:	4 Stroke diesel

6.1 BUILD HISTORY

The Tamar class entered service in 2005 following an extensive design and testing programme which involved both a prototype and pre-production boat, to evaluate the initial design and any subsequent changes.

A total fleet of 27 boats were produced, with production ending in 2013. Being both an afloat boat and slipway boat, the propellers were again heavily tunnelled giving efficiency constraints and a high power demand, similar to the Severn class.

Between the start of development with the prototype boat and the build stage of the pre-production boat, the engine originally trialled had become obsolete, being replaced with a new generation engine. From the prototype boat trials initially requiring a nominal 750hp, and learning from the experience with the Severn class, an engine with a sufficient power margin was selected in an attempt to maintain reliability and achieve the 25-year service life.

Therefore, the engines fitted to the pre-production boat and the rest of the fleet were a pair of new (the first pair in the UK being fitted to the pre-production boat), 18 litre straight 6 twin turbo diesel engines, producing 1000hp at 2300 rpm on a pleasure rating. The engine being fully electronic, with an ECM and MEUI fuel control.

Again, from the previous experience with the Severn, we discussed our design requirements with both the propeller manufacturer and the engine supplier explaining the need for the power margin being as important as top speed to reduce engine load and increase reliability.

As a consequence, the scope of our engine purchase specification increased, not only listing out the key engine components but fixing the power output tolerance to + 3% only.

Also, the engine ECM software was now password protected as standard by the manufacturer, to prevent unauthorised changes to the engine performance parameters.

The pre-production boat was fitted with a 5-blade propeller, giving 24.5 - 25 knts at an engine speed of around 2340 rpm and an engine load of 85 – 87%. This in theory, giving a comfortable 13 – 15% power margin, with the engines sitting well within the 'Preferred rpm range'.

6.2 PERFORMANCE ISSUES

As the boats went into production, at about boat 6, it came to our attention that the boats were now performing extremely well, reports indicating some boats were achieving close to 27 knts.
On further investigation it was also found that the full load rpm had increased to around 2370 rpm and the load was showing 96 - 97%. The engines were now operating above the 'Preferred rpm range'.

How could this be possible, the boats were the same, the propellers were the same and the engines were the same, and more significantly, if the engine was following its overrun curve, at that rpm it should be producing less power not more.

Before this situation could be fully understood, we began to experience catastrophic engine failures (Figure 7).

Figure 7: [Catastrophic failure of a Tamar ALB engine, showing block and con rod damage]

6.3 INVESTIGATION

Examination of the failed engines and discussions with the engine dealer at first, did not provide any answers. However, when the dealer investigated further with the manufacturer, it became apparent that a software change across their entire current marine pleasure rated engine range had been implemented to support an issue raised via the pleasure market.

The manufacturer had modified and extended the overrun curve to provide more power past rated rpm (figure 8).

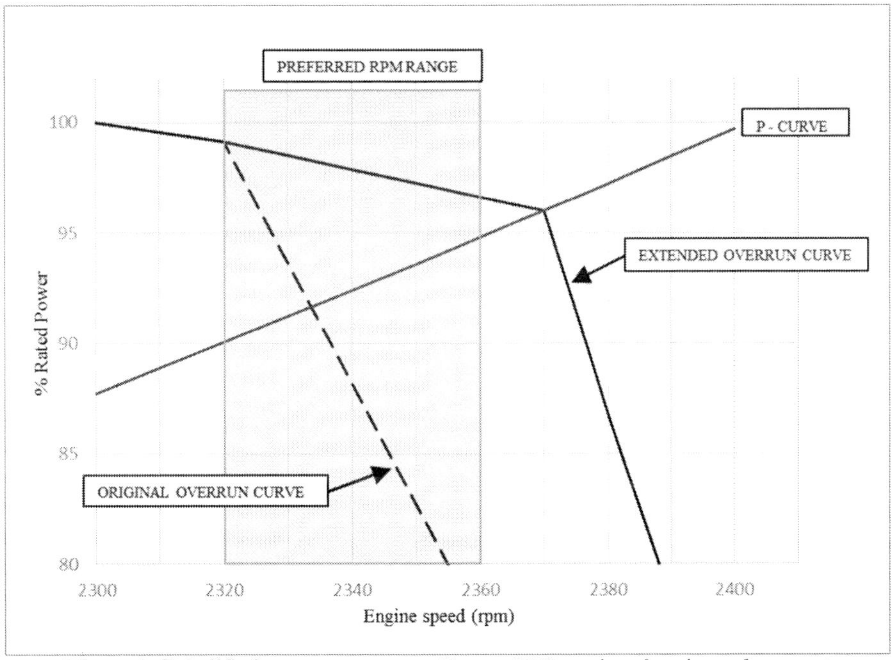

Figure 8: [Modified overrun curve on Tamar ALB engine showing subsequent power demand on the prop curve]

In the pleasure market where speed and every hp counts, having access to all the power the engine can produce is more important than a power reserve. This curve change, maximised engine power available in the 'Preferred rpm range' and made propeller matching easier due to the more gradual power reduction.

However, in our application, with the same propellers fitted, this had led to the propellers absorbing the extra power and the boats gaining extra speed.

This was then slightly masked, once a boat was in service, as extra resistance such as propeller and hull fouling brought the maximum rpm achieved back into the preferred rpm range, and speed back towards 25 knts, except now the engine was producing around 980hp, following the new extended overrun curve.

So where at first glance it would appear from the rpm we were still running with a comfortable power margin, due to a lack of communication from the manufacturer, we were again highly loading the engines.

6.4 THE SOLUTION

To resolve the issue, the engine dealer had to procure blank ECMs for all of the affected engines, load them with the former overrun performance curve and re-commission the engines. Further sea trials confirming they were back within specification.

Our purchase specification was then duly updated to reference the specific software settings for the remaining engine purchases, required to complete the fleet.

6.5 ENGINE LOADING STILL HIGH

Following on from this incident, and although we have had no further significant engine failures, the Tamar class still need careful management of their engine operation.

Subsequent fleet wide modifications, have led to some boats now only just achieving rated rpm at full speed with the engines at 100% load.
The '20's plenty' rule is now very much applicable to the Tamar class as well.

7. CASE STUDY 3 – SEVERN ALB MK2

LOA:	17M
Displacement (t):	42
No. of engines:	2
Propulsion:	5 Blade propeller
Engine configuration:	V10
Capacity (L):	22
Rated power (hp):	1600
Rated rpm:	2450
Cycle:	4 Stroke diesel

7.1 DECISION TO RE-ENGINE

With SLEP still some way off, in 2008 the decision was made to re-engine the Severn class, particularly as the only way to replace some failed engines was through purchasing second hand units and overhauling them.

With the requirement to provide a minimum of 1287hp to ensure the boat would achieve 25 Knts, engines with an initial power output of at least 1350hp and above were reviewed.

With the boats being well into their service life and with current major modifications complete, the engine power requirements were more stable than with a class just entering service. The SLEP itself was not considered an engine performance risk either, as part of the project focus would be on weight reduction.

7.2 ENGINE OPTIONS

With advances in technology since production of the current engine (over 10 years previous), there were initially 5 prime options to choose from. A V10 at 1360 & 1500hp (options A and B – Finally 1600hp as option C), a V10 at 1550hp (option D) and a V12 at 1650 & 1800hp (options E and F) (Figure 9).

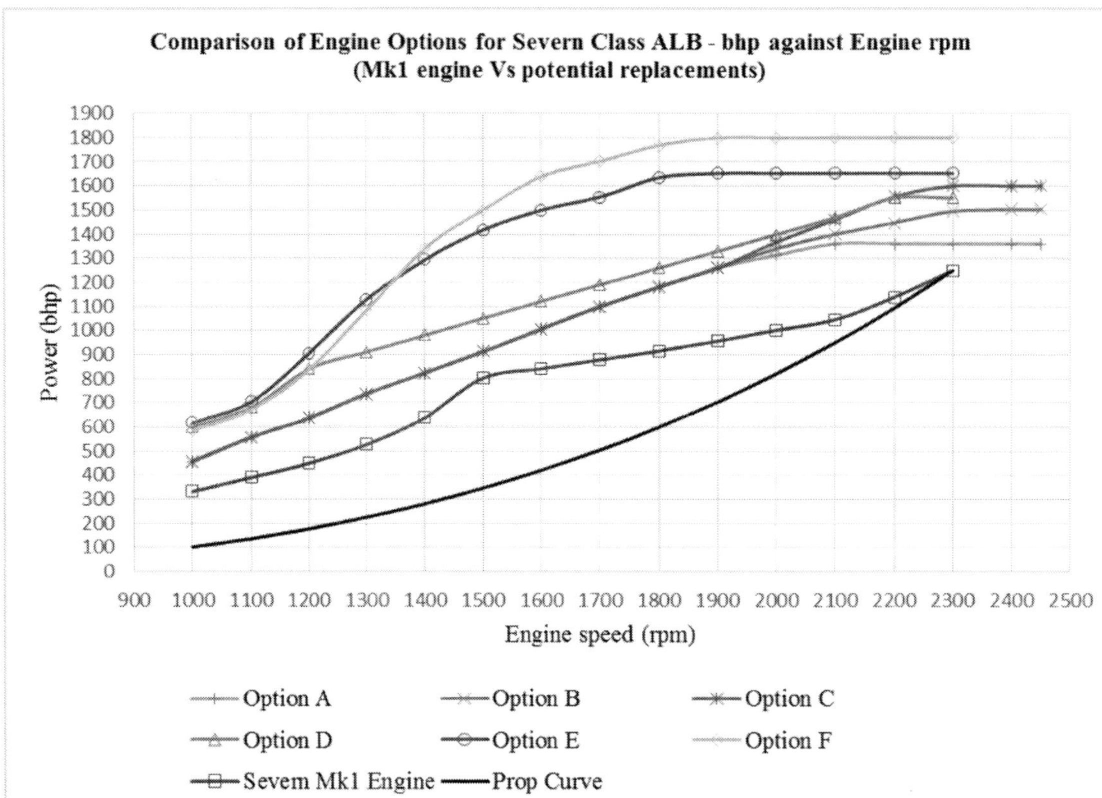

Figure 9: [Power output of potential replacement engines for Severn ALB]

All of the engine options produced significantly more power than required at maximum output and therefore had the potential to operate well below 100% load, however these power outputs were still at a pleasure rating.

All of the options also had significantly improved torque curves over the current engine (Figure 10).

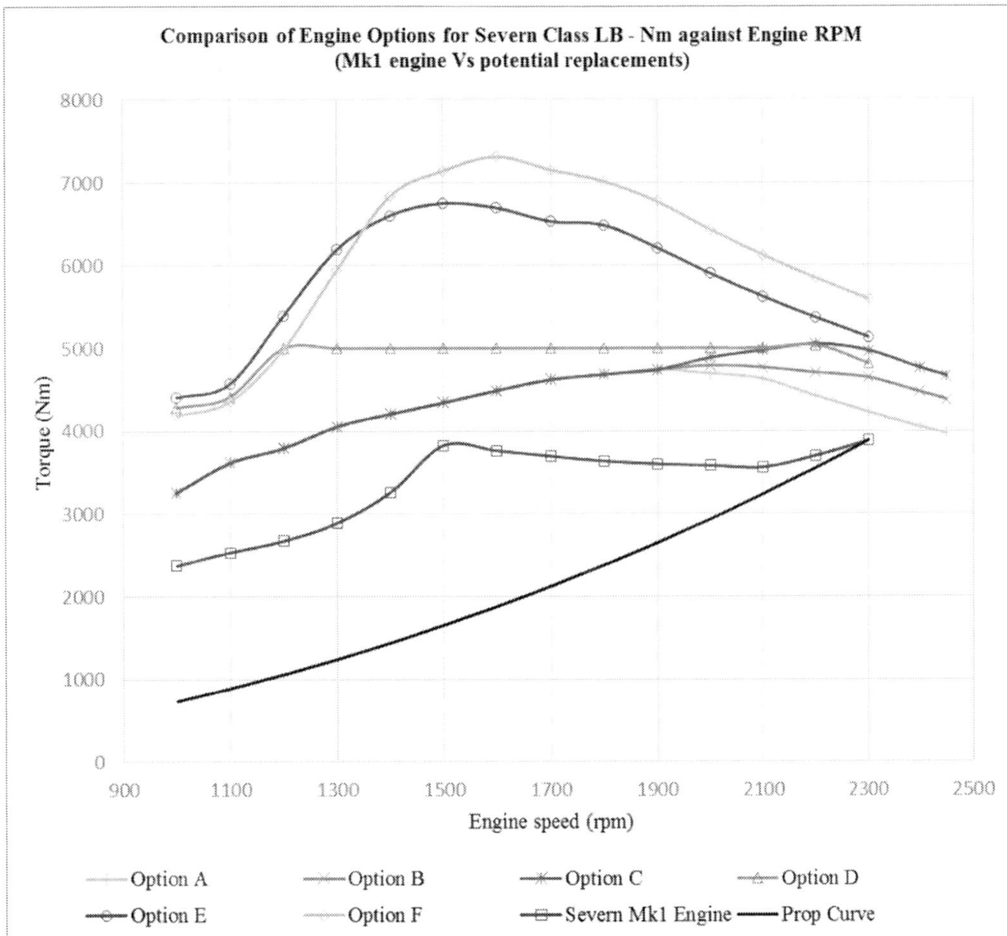

Figure 10: [Torque output of potential replacement engines for Severn ALB]

Whilst the V10 options were of similar weight to, or lighter than the current engine, the V12s although having significantly more power than required, were also significantly heavier, and were discounted on this basis.

Other implications had to be considered as well as providing the additional power to ensure consistent performance, particularly when integrating the new engines with the existing systems that were to remain in place.

The boat was designed with 1250hp engine power output and therefore the rest of the drive train was designed accordingly. The current V-drive gearboxes were also on a pleasure rating and may well have been pushed over their 0.4801 kW/rpm limit if too much power was installed.

As with the V12 engine option, a requirement for a new gearbox to handle the additional power would have incurred additional weight penalties and subsequently required more power, and combined with a heavier engine, any of the additional power benefits may have been lost.

7.3 ENGINE SELECTION

The 20 Litre V10 producing 1500hp (Figure 9 - Option B) was chosen for the trial installation. four of the main reasons for the selection were;

- It had only been in production for 2 years which meant it was relatively new, but well developed
- It was designed specifically for the marine market and not a conversion
- It was lighter and more compact than the current engine
- It had sufficient power reserve

However, another feature of the engine which is key to its use in our application is the fact that the maximum rated rpm is the same as the high idle rpm, there is no overrun curve.

This lack of an overrun curve meant that the propeller could be designed for the required boat speed only, without the need to maintain the engine speed within a 'preferred rpm range' by absorbing sufficient power, giving more opportunity for a greater power margin.

Initial sea trials proved very successful, if not a little too successful with the boat achieving close to 30 Knts, on the first set of propellers. With a new pair of propellers, the boat speed was reduced to the required 25 Knts and a nominal engine loading of 90% achieved. The propeller change required being only a reduction in pitch from that used to resolve the performance issue with the previous engine.

7.4 SOFTWARE ADJUSTMENTS

As the trials progressed it was observed that although the gearboxes were within their limits in the free running condition, they could potentially be overloaded during full speed hard turns and when pushing into big seas.
Where before, boat speed up steep waves was virtually controlled by the lack of the previous engines ability to maintain rpm, and the boat slowed down, the new engines simply increased the fuel being supplied as the load increased, maintaining rpm. The propellers being able to absorb more power without an increase in boat speed.

The engine control software allowed for a torque limit of 4585 Nm to be applied, capping the power output under these conditions protecting the gearbox.

Although the power delivery was described as linear, being fitted with sequential turbo chargers, meant there was a noticeable surge as the second turbo cut in. Standard settings meant this switch point initially coincided with the boat hump speed, leading to cavitation and a loss of grip from the propellers and over speeding of the primary turbo.

Again, adjustments within the engine control software were required and allowed the switching point of the turbo to be moved and the situation avoided.

To gain sufficient confidence in the engine in terms of reliability and performance the trial boat had to undertake an accelerated 500 hr operating period passaging to all of the Severn lifeboat stations around the UK.

7.5 MODEL UPGRADE

As previously mentioned, the typical design life of an engine model is 5 years. With the engine trialled already 2 years old and the installation and evaluation trials taking a further 2 years, the updated model was already available with an increase in power output to 1600hp (Figure 9 – Option C). This giving an increased power margin of 15%.

In order to keep the Severn fleet as current as possible and maximise future support, the engines in the trial boat were replaced with the higher output model.

Discussions with the manufacturer were had to confirm any design changes and their potential implications on performance, the initial review suggesting there shouldn't be any. However further discussions confirmed there were several component updates and software changes.

7.6 ADDITIONAL TRIALS

The new model engines were put through a shortened trials program to confirm performance, again the lack of an overrun curve meant there was no requirement to change or adjust the propellers, the additional power simply remaining in reserve.

Although on paper, there shouldn't have been any changes in boat performance the trials highlighted otherwise. Changes to exhaust manifolds for improved turbo efficiency led to a noticeable delay in acceleration with a cold engine and we were experiencing overspeed issues in rough weather where the propellers were coming out of the water.

We communicated these findings to the engine manufacturer, and whilst we could accommodate the cold delay, the overspeed needed to be investigated. With the engine designers initially adamant that there should not have been any changes between models to affect this, it was only through encouraging them to attend sea trials and reviewing the software parameters together, that a change in the governor response time was identified and re-adjusted.

7.7 ENGINE SPECIFICATION

Following the decision to proceed with the engine replacement for the fleet, the scope of our purchase specification again increased. The specification now not only fixing the power output tolerance to + 3%, but also listing out the key engine components by part number as well as defining the key software parameters and the ECM data set to be used. This was further supported by IMO certification requiring specific components to be fitted in order to meet compliance.

All of the software adjustments made during trials, were also documented in the purchase specification. These changes being completed by either development engineers or dedicated commissioning engineers, ensuring changes were made in a controlled manner.

Following our previous experiences, the decision was made to not give the maintenance software to our service engineers as it was not essential to their role and thus removing any risk of uncontrolled change. Even if they were to have the software, access to engine performance parameters is restricted via 'software keys', the issuing of which is strictly controlled and they would not be able to obtain.

7.8 OVERHAUL PERIOD

It is worth mentioning that whilst the lower power output version of the engine selected, at 1360hp (Figure 9 – option A) was not chosen due to its lack of a power margin, its time between overhauls was significantly higher at 6000 hrs rather than 4000 hrs. With the same power, being demanded from effectively the same engine during normal full speed running, this meant the load on the engine should nominally be the same.

Further discussions with the manufacturer, and following review of our load profile, led to the time between overhaul being extended to 4300 hours. Although this may seem a small gain, this potentially added 2-3 years to the overhaul interval in our application.

7.9 FUEL CONSUMPTION

With additional power, inevitably comes an increase in fuel consumption. With the Severn class carrying 5200 Litres of fuel and an original requirement to achieve 10 hrs at full speed with 10% remaining, this had to be factored into the replacement engine choice. However, through improvements in fuel efficiency with modern engines, it meant we did not incur a significant penalty for the extra power.

Following the engine replacement, there had to be further education with the Severn class crews on the fuel consumption. Although during steady state full speed, the fuel consumption was within limits, the ability for the engines to deliver the extra power in rough weather subsequently meant the fuel consumption would increase and the crews had to be made aware of this, being provided with a load-based fuel curve rather than the traditional boat speed versus fuel burn curve.

Interestingly, modern electronic common rail diesel engines with advances in emission control technology are technically less fuel efficient than they could be. This is due to the multi-phase injection process, injecting to initiate burn, then the main injection to control power followed by a post combustion injection to clean up the exhaust gases. All happening within a fraction of a second, this injection process actually uses more fuel per hp produced than is ultimately possible.

Also of note, is the fact that the engine load percentage displayed on modern engines is not actually measured load, but in fact a percentage of maximum permissible fuel at the current rpm, derived from calculations made by the ECM, and should therefore be treated with caution.

7.10 PROGRESS

To date 32 boats have been re-engined, with 12 remaining to be completed. The boat performance has remained consistent, with some engines now having achieved over 2000 hrs with no failures.

8. CONCLUSIONS

Engine selection is ultimately constrained by the vessel design requirements and the engine models available at the time. However, fully understanding an engine model and its power characteristics, can aid the decision-making process.

Even after selection, engine performance and loadings can only be truly assessed when installed in the intended vessel and put through the required operating conditions.

The importance of being very detailed with the purchase specification of the engine you intend to install should not be underestimated, particularly where large quantities and long time periods are involved allowing production changes to the engine model to occur.

The effects of post installation changes to the engines and other driveline related components in the vessel should also not be underestimated and their implications for both performance and engine reliability need to be fully understood before being implemented.

9. AUTHORS BIOGRAPHY

Spencer Johns is a Senior Engineer at the RNLI. Has worked extensively on propulsion and driveline systems, since joining the RNLI in 2000 and was project manager for the Severn class re-engine programme.

AUTHORS' CONTACT DETAILS

Federico Prini
Marine Offshore and Subsea Technology Group
federico.prini@ncl.ac.uk

Holly J. Phillips
Royal National Lifeboat Institution
hphillips@rnli.org.uk

Vimal Kumar
Lloyd's Register
vimal256@gmail.com

Sean McCartan
EBDIG-IRC, Coventry University
aa2386@coventry.ac.uk

Mohamed Zeid
Caterpillar Propulsion
zeid_mohamed@cat.com

Phill Moxley
Frazer – Nash Consultancy
Phillip.Moxley@babcockinternational.com

Ronald Pelot
Dalhousie University
Ronald.Pelot@Dal.Ca

Jean-Baptiste R.G. Souppez
Southampton Solent University
jean-baptiste.souppez@solent.ac.uk

Albert Nazarov
Albatross Marine Design
an@amdesign.co.th

John W Dalziel
Dalhousie University
john.dalziel@dal.ca

Spencer Johns
Royal National Lifeboat Institution
spencer_johns@rnli.org.uk

Trevor Dobbins
Systems & Technology Research
td@str.eu.com

Thomas Coe
Frazer – Nash Consultanc
Thomas.coe@babcockinternational.com

Hans van der Molen
Koninklijke Nederlandse Redding Maatschappig
he.vandermolen@knrm.nl

The Royal Institution of Naval Architects
8-9 Northumberland Street
London, WC2N 5DA
United Kingdom

ISBN 978-1-5108-8322-2